Environmental Benefits and Costs of Solar Energy

WITHDRAWN

Environmental Benefits and Costs of Solar Energy

WITHDRAWN

Michael D. Yokell
Energy and Resource
Consultants, Inc.

LexingtonBooks
D.C. Heath and Company
Lexington, Massachusetts
Toronto

Library of Congress Cataloging in Publication Data

Yokell, Michael D   1946–
   Environmental benefits and costs of solar energy.

   Includes bibliographical references.
   1. Solar   energy—Environmental   aspects.   2.   Solar   energy—Costs.
I.   Title.
TD195.S64Y64   333.79'23   79–3688
ISBN 0-669-03468-1

Copyright © 1980 by D.C. Heath and Company

Published simultaneously in Canada

Printed in the United States of America

International Standard Book Number: 0-669-03468-1

Library of Congress Catalog Card Number: 79–3688

Contents

List of Figures   vii

List of Tables   viii

Preface   xi

Acknowledgments   xii

Summary   xiii

Chapter 1   **Introduction**   1

1.1 Organization   1
1.2 Objectives   1
1.3 Scope and Limitations   2
1.4 Summary of Methods   5
1.5 Limitations of the Methodology and Data   8

Chapter 2   **Methods**   11

2.1 Method for Projecting Residuals   11
2.2 Method for Projecting Residuals
     Concentrations   34
2.3 Method for Projecting Economic Damages   35

Chapter 3   **Input Assumptions**   39

3.1 Solar-Energy Technologies: Generic
     Descriptions   39
3.2 Solar-Energy Scenarios   43
3.3 GNP, Labor Productivity, and Employment   55

Chapter 4   **Conclusions**   57

4.1 Scenario Analysis   57
4.2 Stand-Alone Analysis   79
4.3 Policy Implications   118
4.4 Further Research   121

Appendix A   **Quality of the Data Sources**   125

Appendix B   **Calculations Used in the AMBIENT and
                BENEFITS Models**   127

                References   131

                Index   139

                About the Author   143

List of Figures

1-1	SEAS Flow Chart	7
2-1	ESNS Network	16
2-2	ESNS Solar Network	18
2-3	INFORUM Flow Chart	19
2-4	Energy Scenario Strategy	24
3-1	Historic Rate of Per-Capita Energy Use	44
3-2	Historic Ratio of Total Energy Demand to GNP	45
3-3	Residential and Commercial Per-Capita Energy Use	46
3-4	Industrial Per-Capita Energy Use	46
3-5	Transportation Per-Capita Energy Use	47
3-6	Electricity Per-Capita Energy Use	47
3-7	Energy Sources, Base Cases	49
4-1	Federal Regions of the United States	83
4-2	Cumulative Life-Cycle Particulate Emissions, 1985 Data	99
4-3	Cumulative Life-Cycle Sulfur Oxide Emissions, 1985 Data	100
4-4	Cumulative Life-Cycle Nitrogen Oxide Emissions, 1985 Data	101
4-5	Cumulative Life-Cycle Hydrocarbon Emissions, 1985 Data	102
4-6	Cumulative Life-Cycle Carbon Monoxide Emissions, 1985 Data	103
4-7	Cumulative Life-Cycle BOD Emissions, 1985 Data	104
4-8	Cumulative Life-Cycle Suspended-Solids Emissions, 1985 Data	105
4-9	Solar-Energy Deployment as a Function of Time	119
4-10	Rate of Growth of Solar-Energy Deployment as a Function of Time	121

List of Tables

1–1	Residuals Computed and Analyzed	3
2–1	Major Recent National Energy Scenarios	23
2–2	Macroeconomic Assumptions for the Base-Case Scenarios	25
2–3	Census Region Solar Demands for Year 2000	28
2–4	RESGEN Solar Data Base	32
2–5	BENEFITS Damage Functions	37
3–1	Summary of Solar Energy in Base Cases by Year 2000, Quads	48
3–2	Base Case (DPR 1, SERI 1) Mid Price: U.S. Energy Demand and Supply for Year 2000, Quads	50
3–3	Maximum-Practical Case (DPR 4, SERI 2): U.S. Energy Demand and Supply for Year 2000, Quads	51
3–4	Maximum-Feasible Case (DPR 5, SERI 3): U.S. Energy Demand and Supply for Year 2000, Quads	52
3–5	Summary of Solar Energy in Maximum-Practical and Maximum-Feasible Cases, Year 2000, Quads	53
3–6	Assumed Regional Allocation of Solar Energy for All Scenarios, Year 2000	54
4–1	Comparison of the Base, Maximum-Practical, and Maximum-Feasible Cases for the Year 2000: National	59
4–2	Sectors Causing Changes in National Residuals Outputs between the Maximum-Practical Scenario and the Base Case for the Year 2000	60
4–3	Comparison of High Environmental Control Technology (Scenario 1H) and Maximum-Practical and Maximum-Feasible Scenarios (2L and 3L) with Low Control (1L) for the Year 2000: National	65
4–4	Relative Effects of Pollution Control on Scenarios in the Year 2000: National	66
4–5	Comparison of Discounted Expected Environmental	

	Damages, 1975 to 2000: Alternative Scenarios, Values of Life, Discount Rates	69
4–6	Environmental Benefits of Solar Energy under Mid-Range Assumptions	72
4–7	Cumulative Expected Damages, 1975 to 2000: Alternative Scenarios, Alternative Discount Rates (Value of Life = $30,000)	73
4–8	Cumulative Expected Damages, 1975 to 2000: Alternative Scenarios, Alternative Discount Rates (Value of Life = $100,000)	74
4–9	Cumulative Expected Damages, 1975 to 2000: Alternative Scenarios, Alternative Discount Rates (Value of Life = $200,000)	75
4–10	Cumulative Expected Damages, 1975 to 2000: Alternative Scenarios, Alternative Discount Rates (Value of Life = $300,000)	76
4–11	Sectors Causing Changes in Net Discounted Environmental Damages	77
4–12	Cumulative Expected Damages and Benefits, 1975 to 2000: Alternative Scenarios by Pollutant	78
4–13	National Expected Air Pollution Damages for Alternative Scenarios	80
4–14	Comparison of Air Damages by Federal Region: Year 2000	81
4–15	Comparison of Water Damages by Federal Region: Year 2000	82
4–16	Annual Emissions of Particulates from Solar-Energy Facilities (SEAS Model)	87
4–17	Annual Emissions of Sulphur Oxides from Solar-Energy Facilities (SEAS Model)	87
4–18	Annual Emissions of Nitrogen Oxides from Solar-Energy Facilities (SEAS Model)	88
4–19	Annual Emissions of Carbon Monoxide from Solar-Energy Facilities (SEAS Model)	89
4–20	Annual Emissions of Hydrocarbons from Solar-Energy Facilities (SEAS Model)	89

4–21	Annual Water Emissions in Biological Oxygen Demand from Solar-Energy Facilities (SEAS Model)	90
4–22	Annual Water Emissions of Suspended Solids from Solar-Energy Facilities (SEAS Model)	91
4–23	Annual Emissions of Particulates from Conventional Energy Facilities (SEAS Model)	91
4–24	Annual Emissions of Sulphur Oxides from Conventional Energy Facilities (SEAS Model)	92
4–25	Annual Emissions of Nitrogen Oxides from Conventional Energy Facilities (SEAS Model)	92
4–26	Annual Emissions of Carbon Monoxide from Conventional Energy Facilities (SEAS Model)	93
4–27	Annual Emissions of Hydrocarbons from Conventional Energy Facilities (SEAS Model)	93
4–28	Annual Water Emissions in Biological Oxygen Demand from Conventional Energy Facilities (SEAS Model)	94
4–29	Annual Water Emissions in Suspended Solids from Conventional Energy Facilities (SEAS Model)	94
4–30	Modeling Assumptions: SEAS Stand-Alone Residuals Projections	95
4–31	Annual Emissions of Particulates from Coal Acquisition and Delivery	97
4–32	Annual Emissions of Sulfur Oxides from Coal Acquisition and Delivery	97
4–33	Annual Emissions of Nitrogen Oxides from Coal Acquisition and Delivery	98
4–34	Active Solar Heating Systems: Candidate Working Fluid Characteristics	107
4–35	Heavy Metal Content in Anaerobic Digestor Sludge	115
4–36	Trace Metal Content of Wood-Derived Ash	117
A–1	Quality of the Solar-Residuals Coefficients Data	125
A–2	Quality of the Energy Investment Module Data for 1985	126

Preface

This book is the result of nearly two years of effort conducted mainly while the author was a staff member at the Solar Energy Research Institute (SERI) in Golden, Colorado, Department of Energy (DOE) laboratory operated by the Midwest Research Institute (MRI). The contents of this publication do not represent the views of SERI, MRI, or DOE, and no warranty, express or implied, is made by those institutions with respect to the accuracy or completeness of information contained herein. Due to space limitations, methodological details, calculations, and input data have been omitted. For example, a detailed series of process-analysis-type calculations were made to check the accuracy of the input-output results. These calculations generally verified and validated the input-output work and are not reported here.[a] Another example concerns the regional distribution of environmental costs and benefits of solar energy. These results turned out to be uninteresting and were omitted accordingly. The main conclusions, however, are reported.

In a study of this magnitude, regrettably, many details cannot be published in book form due to space limitations. Interested readers requiring specific information can obtain it by writing to the author at Energy and Resource Consultants, Inc., P.O. Drawer O, Boulder, Colorado, 80306.

---

[a] Data sources used in the process-analysis calculations included Ballou (1978), Caputo (1977), Energy and Environmental Analysis (1977), Lawrence (1979), MITRE (1977, 1977a, 1977b), Neenan (1979), Sittig (1975), and U.S. Environmental Protection Agency (1977, 1978).

Acknowledgments

Several SERI staff members contributed to this book. Richard Caputo, senior engineer, was on leave from the Jet Propulsion Laboratory, Pasadena, California. He provided the national scenarios used to run the Strategic Environmental Assessment System (SEAS) model. Dennis Horgan, staff engineer, provided the generic design for the photovoltaic systems described in section 3.1. Kathryn Lawrence, staff biologist, provided many critical insights and wrote the majority of section 4.2. Al Shaheen provided research assistance to the project. Harit Trivedi, staff statistician, provided much of the data described in section 1.3.

Members of the Department of Energy's (DOE) Technology Assessment of Solar Energy Program were helpful in providing some of the capital-cost data and some of the residuals data. Individuals who were helpful include Robert Blaunstein, DOE program manager, John Altseimer, principal investigator, Bernard Neenan and Vince Gutschick of Los Alamos Scientific Laboratory, and their subcontractors at other national laboratories and private firms.

Roger Shull of DOE was especially helpful in arranging for the use of the SEAS model and providing both computer time and funding for those portions of this work that were permanently incorporated into the model.

The work could not have been carried out without the very cooperative and able assistance of the MITRE Corporation and William Watson, contractors; and CONSAD Research Corporation, Control Data Corporation, and International Research and Technology, Inc., subcontractors. Individuals from these companies who worked on the project include Debbie Elcock, Richard Kalagher, Rob Kline, and Andy Lawrence of MITRE; Nazir Dossani and Bill Weygandt of CONSAD; Raj Shah and Brad Wing of Control Data; and Marc Narkus-Kramer, Marylynn Placett, and Brian Rennex of International Research and Technology.

I would also like to thank the approximately twenty-five individuals who attended the June 1978 workshop, "Economic Measurement of Energy-Related Environmental Damages," that was part of this effort. Many useful insights were contributed by the participants.

Summary

The principal conclusion of this book is that, on a national basis, rapid deployment of solar energy would make a significant net contribution to environmental quality during the 1975–2000 period. Using significant but practical amounts of solar energy during this period would reduce nearly all air and water pollutants; the principal exception is particulates emitted to the atmosphere. Reductions in atmospheric emissions of sulfur oxides are projected to produce major environmental benefits from solar energy, far outweighing increased particulate damages. These benefits are portrayed in table 4–5 as a function of the real discount rate and the value of human life assumed in the calculations. Particulate damages stem from the use of wood stoves and the increase in the size of the stone-and-clay-products industry required to produce some solar systems. Agricultural runoff from silvicultural farming practices may also be an environmental problem. Although the use of solar energy results in major environmental benefits, more directly imposed environmental controls would result in much larger environmental improvements than a switch to solar energy over the 1975–2000 period. These conclusions cannot be extended easily to federal regions. All regions benefit to some extent, however. The reduction in CO_2 emissions achievable by the year 2000 under even the maximum-feasible solar-energy scenario is negligible compared with the global atmospheric inventory of CO_2.

The second major conclusion is that, on a life-cycle basis, nearly all the solar technologies modeled are more environmentally benign than their conventional counterparts per Btu of energy delivered per year. In addition, the nontransportation-sector emissions associated with a centralized photovoltaic system are somewhat greater than those associated with the decentralized photovoltaic-deployment mode. The construction impacts per Btu per year of individual solar-energy systems modeled, however, uniformly exceed those of conventional systems considered. This has profound policy implications. By maximizing the rate of solar deployment in the present period, we hasten the time when the percentage of energy supplied by solar systems has reached a steady state; thus, we hasten the time when society begins to reap fully the potential environmental benefits available through the deployment of solar energy. On the other hand, during a period of rapid growth in the deployment of solar-energy systems, the environmental benefits available are temporarily reduced from what they would be under slower deployment conditions.

If President Carter's Domestic Policy Council's solar-deployment goals for the year 2000 are to be met and the generation that is living as the year 2000 approaches is not to be environmentally shortchanged at the expense of post-2000 generations, then U.S. policy must favor more rapid growth in the deployment of solar-energy technologies in the early part of the 1975–2000 period with a tapering off in the rate of growth toward the latter part of that period.

A number of other policy implications emerge from the study. An important conclusion is that solar-energy technologies could be useful tools for environmental control in selected regions. If the solar technologies whose operating phases are most environmentally benign (for example, solar heating and cooling of buildings and photovoltaics) were deployed in areas where the current technologies are based on oil or coal combustion and, at the same time, if the construction of the solar-energy facilities could be located outside the region, then an environmental transfer payment from one region to another would be effected. This conceivably could be done as a matter of deliberate policy in constricted air sheds, such as the South Coast Air Basin in southern California.

Not all solar technologies are environmentally benign, however, even in the operating phase. Wood stoves are a significant contributor to particulate emissions under the Domestic Policy Council's scenarios. This has important implications for public policy. Either a cost-effective technology to control particulate emissions must be found or a limit must be set on the use of wood stoves in regions with air-quality problems or meteorological conditions that would be adverse to large particulate emissions.

Silvicultural farming is the other solar technology with major direct environmental effects associated with the operating phase. Here the problem is agricultural runoff. The solution appears to be careful monitoring of runoff combined with further research in runoff-minimizing farm practices.

The emission of particulates from the stone-and-clay industry, which is stimulated by the construction requirements for a number of the solar technologies, is the major indirect environmental effect of the group of solar technologies. Since the stone-and-clay-products industry is already regulated under the Environmental Protection Agency's New Source Performance Standards, both research and policy must strive to reduce the quantities of stone and clay products used in the construction of solar facilities.

This conclusion is actually a general one: to the extent possible, research and policy should strive to reduce the materials intensity of the solar technologies. Although this is the object of much engineering research in the solar field, the environmental effects of materials-intensive production technologies imply that materials-reducing research and policy should be carried beyond the point that would be optimal from an engineering-economics perspective.

The principal objectives of the research were to quantify the environmental benefits and costs of six generic solar-energy systems: solar heating and cooling of buildings, agricultural and industrial process heat, solar-thermal electricity, photovoltaics, wind-energy-conversion systems, and biomass for energy.

In this research two separate kinds of environmental analyses were carried out. First, and most important, was an analysis based on national energy scenarios in which solar energy is substituted for conventional sources of energy between 1975 and 2000. Three scenarios were examined: a base case, a maximum-practical solar scenario, and a maximum-feasible solar scenario. Each of these scenarios was studied under both strict and lax environmental controls. The second analysis compared individual solar-energy systems with conventional technologies on a stand-alone basis, according to environmental impact per unit of delivered energy. This analysis was carried out at the national level for the years 1975, 1985, and 2000. The data for 1985 were cross-checked using process analysis.

The approach used to estimate the operating or direct environmental impacts of solar technologies involved an extensive literature search coupled with personal interviews with experts in individual solar-energy technologies. Indirect effects were evaluated using the Strategic Environmental Assessment System (SEAS), a computer model based on input-output techniques. This approach was supplemented by manual calculations based on process analysis. A method for the economic evaluation of these environmental impacts was developed using the AMBIENT and BENEFITS computer programs originated at Resources for the Future, Inc.

Several methodological improvements in this work are possible. For example, there will always be a need for site-specific environmental analysis of large, proposed energy facilities. Also, specific regional analyses, conducted in response to bona fide governmental policy options that will affect the region in question, may be useful. One of the great difficulties of this study was attempting to quantify economically the benefits and costs of solar energy. The principal difficulty was the lack of scientifically acceptable environmental damage functions. Environmental-economics research is an area that deserves greater attention. Improvements would enhance greatly the confidence that could be placed in the results of the benefit-cost analysis.

1 Introduction

1.1 Organization

This book is divided into four chapters. The first describes the objectives of the research, provides a summary of the methods used to accomplish these objectives, and discusses the limitations of the work that derive from limited scope and deficiencies in methodology and data. The second chapter describes in detail the methods that were used in this analysis. The third chapter is a summary of the assumptions that were made. Generic solar-energy technologies are described, the scenarios that were analyzed are discussed, and other assumptions are detailed. The fourth chapter presents the conclusions of this study. Brief sections on the policy implications of the results and useful additional research are included.

1.2 Objectives

The principal objective of this research was to quantify, in physical terms, the environmental benefits and costs of deploying six generic solar-energy systems: solar heating and cooling of buildings (SHACOB), agricultural and industrial process heat (AIPH), solar-thermal electricity, photovoltaics (PV), wind energy conversion systems (WECS), and biomass for energy. A secondary objective was to measure these benefits and costs in economic terms.

Any technology has environmental costs associated with it, stemming from pollution-induced damages to human health, crops, buildings, and quality of life. Conventional energy technologies are particularly intensive in the use of existing environmental amenities. The environmental benefits or costs of any energy technology that would substitute for conventional technologies should be evaluated comparatively. Thus the environmental benefits of solar-energy systems have been measured by the environmental costs engendered minus the environmental costs of the conventional systems supplanted.

The environmental effects of any technology have been divided into two categories: those that result directly from the operation and maintenance of the technology once in place (*direct effects*) and those that result from the manufacture of the necessary materials and their assembly into a

1

usable system (*indirect effects*). In general, with the exception of some bio-mass energy systems, the direct environmental effects of solar-energy systems are small. Thus, most of this investigation has focused on the indirect effects of solar-energy systems.

Environmental benefits and costs are a subset of social benefits and costs. Social benefits and costs of proposed production processes, such as solar-energy technologies, need to be carefully evaluated. The principal mechanism for selecting technologies in our society, the private market-place, does not adequately account for all social benefits and costs. This omission, one reason for what economists call *market failure*, has engen-dered an extensive body of literature, much of it summarized in Baumol and Oates (1975).

The possibility that solar-energy technologies might have significant environmental benefits not ordinarily assessed in the marketplace (*positive externalities*) is particularly important since solar-energy systems generally are not yet competitive with conventional technologies in the marketplace (Flaim et al. 1978). The possibility that environmental benefits from solar-energy technologies might make them socially, if not economically, com-petitive was one of the principal motivations behind this book.

1.3 Scope and Limitations

The focus of this book is twofold. First, the environmental effects of solar and conventional energy technologies are analyzed. Second, an experimen-tal methodology is applied to analyze the environmental benefits and costs of solar and conventional energy scenarios. Six generic solar-energy tech-nologies (subdivided into sixteen different options) were studied and com-pared with conventional energy technologies. These sixteen options do not cover the universe of possible solar-energy systems. Also, since this was a national policy study, siting of particular energy systems was very general. Consequently, those environmental effects that are highly site specific were not analyzed. The study should provide useful input to national energy and environmental policy decisions that cannot be concerned with the details of site-specific analysis. Not all possible environmental effects were consid-ered. In particular, the environmental and safety risks of low-probability, high-impact events, such as serious nuclear accidents, were not treated. Two solar-energy technologies, ocean thermal-energy conversion (OTEC) and solar-power–satellite stations (SPSS) were not considered since it is unlikely they would be commercialized by the year 2000 when the planning horizon for this book concludes (Flaim et al. 1978).

Not all possible options have been examined within each technology. For example, no systems that convert biomass to liquid fuels were examined. Moreover, within each option only one design was chosen for analysis, though many are often technically feasible. For example, cadmium-sulfide-photovoltaic cells were assumed to be used in both centralized and decentralized configurations, though other choices, such as silicon or gallium arsenide cells were possible.

Many of the environmental effects of energy technologies are site specific. Some, like the effect of construction activities and some maintenance operations, are so specific that they can only be analyzed with intimate knowledge of the surrounding ecosystems. Others, like the impact of carbon-dioxide emissions on global climate, have little to do with site specification. Between these extremes are a number of environmental effects that can be analyzed regionally but that could be treated more thoroughly at specific sites if sufficient research resources are available. Among these effects are pollutant emissions, and it is these emissions on which this book focuses principally.

Not all possible residuals have been considered. For example, occupational deaths were not estimated. A list of those residuals that have been computed and analyzed appears as table 1–1. Additionally, some pollutants specific to certain solar-energy systems, such as cadmium emissions, were considered.

Finally, low-probability, high-impact events, such as those that might be associated with a nuclear accident, were not considered. This does not necessarily mean that nuclear power is environmentally benign and safe. Rather, the statistical data available simply do not permit consideration of potential nuclear accidents within the same framework as those of other energy technologies. The indirect environmental impacts of nuclear power plants were analyzed, however.

The damages associated with pollution emissions are significantly more difficult to project than the emissions themselves. In the economic evaluation of environmental damages, only the following pollutants are treated.

Table 1–1
Residuals Computed and Analyzed

Residual	Residuals Computed	Economic Analysis Performed
1 Particulates	X	X
2 Sulfur oxides	X	X
3 Nitrogen oxides	X	X
4 Hydrocarbons	X	X

Table 1-1 continued

Residual	Residuals Computed	Economic Analysis Performed
5 Carbon monoxide	X	X
6 Photochemical oxidants		
7 Other gases and mists		
8 Odors		
9 Biochemical oxygen demand	X	X
10 Chemical oxygen demand	X	X
11 Total organic carbon		
12 Suspended solids	X	X
13 Dissolved solids	X	X
14 Nutrients		
15 Acids		X
16 Bases		
17 Oil and greases		X
18 Surfactants		
19 Bacteria		
20 Waste water		
21 Thermal loading		
22 Combustible solid waste		
23 Noncombustible solid waste	X	
24 Bulky waste		
25 Hazardous waste		X
26 Mining waste		
27 Industrial sludges	X	
28 Sewage sludge	X	
29 Herbicides		
30 Insecticides		
31 Fungicides		
32 Miscellaneous pesticides		
33 Radionuclides, air		
34 Radionuclides, water		
35 Radionuclides, deep burial		
36 Radionuclides, shallow burial		
37 Water use	X	
38 Land use (plant lifetime)	X	
39 Land use per year	X	
40 Injuries	X	
41 Man-days lost	X	
42 Operational labor		

For air, total suspended particulates, sulfur oxides, nitrogen oxides, carbon monoxide. and hydrocarbons were taken into account. For water, damage estimates are based on acidity, biological oxygen demand (BOD), chemical oxygen demand (COD), dissolved solids, heavy metals, nitrates. oil and grease, and suspended solids.

1.4 Summary of Methods

The approach used to estimate the direct environmental impacts of solar technologies involved an extensive literature search, coupled with personal interviews with experts in individual solar energy technologies. Indirect effects were evaluated using the Strategic Environmental Assessment System (SEAS), a computer model based on input-output techniques. This approach was supplemented by manual calculations based on process analysis. A method for the economic evaluation of these environmental impacts was developed using the AMBIENT and BENEFITS computer programs developed at Resources for the Future, Inc. (RFF).

Two separate kinds of environmental analyses were performed. First was an analysis based on national-energy scenarios in which solar energy substitutes for conventional sources of energy between 1975 and 2000. Three scenarios were examined: a base case, a maximum-practical solar scenario, and a maximum-feasible solar scenario. Each of these scenarios was studied under both strict and lax environmental controls. Impacts were disaggregated geographically to the county level by computer, though they are reported here only at the federal region level. The scenario-based analyses allowed projection of the rich variety of changes in emissions which can take place at various locations and times as solar-energy sources are continuously substituted for conventional sources of energy.

One of the difficulties with a scenario-based analysis that stops at the year 2000 is that the percentage of solar energy used by the nation is most likely to be growing rapidly at that time. Since the indirect environmental effects of solar-energy systems predominate over the direct effects (while the converse is true for at least some conventional energy sources) and the indirect effects occur at the beginning of a solar system's life cycle, solar-energy sources might be more environmentally damaging during this period than they would in a *steady-state economy* in which the fraction of energy contributed by solar sources is constant. As a partial solution to this problem, a second analysis was performed in which individual solar-energy systems were compared with conventional technologies on a *stand-alone basis,* according to environmental impact per unit of delivered energy. This analysis was done using projected data on national production relations and emissions for the years 1975, 1985, and 2000. The data for 1985 were cross-checked using process analysis.

Following projection of the environmental emissions attributable to the various solar-energy systems, it was desirable to estimate the economic value of the associated benefits and costs in order to have a common basis for comparing the economic and environmental benefits and costs of alternative energy policy choices. An experimental analysis of the economic benefits and costs attributable to solar-energy scenarios was carried out using a simulation program devised at RFF. These benefits and costs were calculated only for complete scenarios and not on a technology-by-technology basis. The reason was that the environmental damages attributable to a particular technology and emissions stream depend on when and where the technology is located, among other factors. For example, the human health costs of a coal-fired electric generation plant are much greater if the plant is located nearby and upwind of a major city than if located remote from human habitation. In general, damages are an increasing function of emissions for any given pollutant. Thus, the concept of the environmental benefits and costs of solar-energy systems is most meaningful when applied to an entire economy.

The SEAS and the AMBIENT/BENEFITS model are described briefly here. A more complete description of each is presented in chapter 2.

SEAS is a system of interlinked computer models of the energy system, economy, and environment. It is a projecting model rather than an optimizing model. A flow chart of SEAS is presented in figure 1–1. The heart of SEAS is a 200-sector input-output model of the economy, called INFO-RUM. An input-output model is organized around a table of input-output coefficients, or *A matrix*, that specifies the dollar amount of output from each sector of the economy required to produce a dollar's worth of each sector's output. *Final demands* are the amounts of each sector's output that are delivered to consumers. Final demands are synonomous with gross national product (GNP). GNP is defined as the sum of personal consumption expenditures, gross private investment, government expenditures, and net exports. *Intermediate demands* are the amounts required to produce final demands, and *total demands* are the sum of the two.

SEAS works basically as follows. First, the structure of the U.S. energy system is detailed in a *spaghetti-bowl diagram* in which energy end uses are traced back to their energy supply sources. Next, energy demand and supply are provided exogenously, along with data on construction lead times for various types of energy facilities. This determines the energy-related investment required in each year to meet the exogenously specified demand levels. Third, disaggregate data on the dollar value of materials requirements for each type of energy facility are provided. These are used to determine investment by energy sector. These estimates modify final demands in INFO-RUM. (An identical procedure is used to model the investment impact of pollution abatement expenditures.) Fourth, pollution abatement expendi-

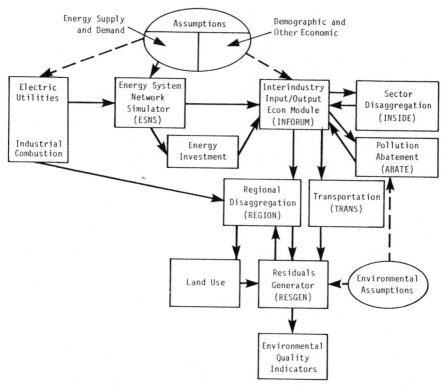

Source: MITRE Corp. 1978a, *A Short Course on the SEAS Model.* McLean, Va., May 1978.
Reprinted with permission.

Figure 1-1. SEAS Flow Chart

tures also impact INFORUM final demands and A matrix coefficients. After the national economic model is run, it is disaggregated both by region and more detailed sectors. The sectors for disaggregation are chosen for environmental importance. Regional disaggregation is carried out at the county level. Finally, a series of environmental models simulate pollution generated at the county level from each of the disaggregated sectors. County-level data can be aggregated to states, federal regions, standard metropolitan statistical areas (SMSAs), and so forth.

A thorough environmental analysis of solar-energy technologies cannot stop at pollutant emissions. It should continue to analyze the environmental stresses and damages associated with the projected emissions. For this analysis, the AMBIENT/BENEFITS system was used.

AMBIENT is a simple model that projects the concentrations of pollutants in the future based on historical concentration data and projected emissions data. Forecasted population levels are then used to project pollution exposures per capita.

The damages associated with projected levels of pollution are critical variables in the decision-making process. The BENEFITS model was used in an experimental mode to estimate the dollar value of environmental damages that would occur in each scenario. The BENEFITS model is based on regionalized damage functions for air and water pollutants and on a control total for national pollution damages in 1975. These damage functions and the control total rely on very incomplete data. Consequently the results from the BENEFITS model should be treated as indicating general trends rather than as hard data.

1.5 Limitations of the Methodology and Data

The limitations of the study that result from the methodology and data utilized are described here. The first limitation is that the calculation of direct environmental emissions attributable to solar technologies is based on data in available literature. For a number of technologies, particularly biomass technologies, little prior research has been done. More experimental knowledge would be beneficial.

Analysis of indirect emissions is based on the input-output analysis embedded in the SEAS/INFORUM model. Input-output analysis suffers from the following weaknesses: (1) it assumes a linear relationship between inputs and outputs; (2) it assumes constant returns to scale in all production processes; (3) input-output data take several years to collect and can be outdated by the time they are processed; (4) as prices and technology change, the structure of the economy changes, which makes it difficult to project these changes using input-output analysis; and (5) aggregation errors are inherent in the use of input-output models—even a 200-sector model is highly aggregated in comparison to basic engineering processes.

The SEAS model has two major weaknesses in addition to those associated with any input-output model. These are the use of a *top-down* regionalization procedure and the absence of any optimizing capability. In SEAS, point-source emissions are based on emission coefficients multiplied by the regional output of either economic or energy sectors. In some cases, regional residual coefficients are utilized; these are then multiplied by regional energy or economic output. Regional energy outputs can be specified as input to the SEAS model, or computer routines are available as SEAS options to locate industrial-combustion and electric-generation facilities. These siting routines are a potential source of weakness in the regional-emissions forecast. INFORUM economic sectors were also regionalized using one of the SEAS modules called REGION. REGION is based on population forecasts and OBERS economic forecasts (U.S. Department of Commerce 1974). Regional disaggregation of this sort is inherently weaker than bottom-up aggregation of regional forecasts to the national level.

SEAS is a logically consistent way of making emission projections. It is not an optimizing model and does not have constraints built in as a linear programming model does. Therefore, the projections that can be made with SEAS are no better than the scenarios that are entered; that is, the model will not correct for inconsistencies in the scenarios. For example, a scenario might be supplied as input in which coal mining in Region VIII requires more than the available water in that region, yet SEAS would not consider this constraint. Thus careful creation of scenarios is extremely important to the modeling effort.

The AMBIENT model, used to project ambient pollution concentrations, suffers from several deficiencies. AMBIENT is unable to account for changes in pollution transport phenomena over time. Also it is limited to only five pollutants (TSP, SO_2, NO_x, CO, and HC). AMBIENT's source-receptor transfer coefficients, which are used to incorporate meteorology and transport processes, are based on very limited data. The same set is used in every region.

The AMBIENT system produces a single annual average pollution concentration value for each pollutant near the population center, rather than a distribution of concentration values for alternative time intervals and over geographic areas. Another problem with AMBIENT is the use of the prevalence, duration, intensity (PDI) index of water pollution. This index relies heavily on subjective judgments and is not universally accepted or in widespread use.

The BENEFITS module, used to calculate economic damages associated with projected pollution concentrations, has five major weaknesses.

1. To fully account for environmental effects, differences in net residual production between solar-energy scenarios should be converted to impacts on human health, property, and ecological systems. However, the dose-response functions used in BENEFITS are implicit economic-damage functions, rather than explicit physical dose-response functions.

2. Per-capita economic-damage functions are concave upward and have slopes (relative to the median exposure level) that are equal to the slopes of a few existing empirical functions, or they have slopes based on judgment. These damage functions are highly generalized and often lack theoretical and empirical justification.

3. The BENEFITS model uses population as a proxy for all values at risk even though other values, such as materials and residential properties, are also at risk.

4. The highest value of life used in the BENEFITS model is $300,000 (1972 dollars). This may substantially underestimate the willingness to pay

for reductions in the probability of death. Some existing studies show values as high as $1 million per life (Fisher 1979). This conservative estimate of the value of life biases downward the environmental benefits of solar energy estimated using the BENEFITS model.

5. National environmental-damage estimates incorporated in the BENEFITS model are based on fragmentary regional studies extrapolated to the national level. In many cases, insufficient numbers of regions were studied to allow confidence in the extrapolation procedure.

This chapter is based in part on a workshop held in conjunction with this work and reported in the *Workshop Summary* (Yokell 1978). For a more thorough understanding of the methodological limitations, a careful study of chapter 2 is necessary; for a better understanding of the deficiencies of the input data utilized, chapter 3 should be examined.

2 Methods

2.1 Methods for Projecting Residuals

2.1.1 Introduction

Two general methods were used to project the residuals indirectly associated with individual solar-energy technologies: process analysis and input-output analysis. Process analysis was done only for 1985. Input-output analysis was carried out for the years between 1975 and 2000. Results from the two methods were compared and, to the extent possible, reconciled. Due to space limitations, only the input-output results are reported here.

In process analysis, each step in the manufacturing process for a particular solar technology is analyzed for its potential residuals contribution. In theory, each step in the manufacturing process for each of the important inputs to the solar manufacturing process could be examined similarly. In this study, however, only the first step in the manufacturing process was analyzed since more detailed process analysis would have been impractical under available resource constraints.

The emphasis was on input-output (I/O) analysis rather than process analysis for several reasons. First, I/O analysis is comprehensive in that it traces the complete chain of sectoral impacts resulting from an energy-induced change in the economy. Second, I/O analysis is a self-consistent method of projecting changes in nonsolar-energy technology that ultimately affects the residuals produced by any given solar technology. That is, as the economy changes, the residuals associated with using that economy to produce solar systems change. I/O analysis makes it possible to trace this process. Finally, the SEAS model, built around an I/O model, provided a large integrated set of data bases that was extremely useful in the projection of pollutant emissions associated with solar-energy technologies and scenarios.

Input-output analysis is an automated version of process analysis, carried out in less detail but for more steps in the processing chain. I/O analysis begins by dividing the economy into representative *sectors,* or industries, the fineness of division depending on the purposes for which the analysis is being used and the resources available to the investigator. Typical sectors in the 200-sector model used in this analysis are steel, stone and clay products, and nonferrous casting. I/O analysis uses data on dollar purchases of each sector's output to create a system of simultaneous linear equations that

11

describe the transactions among all the sectors of the economy and final consumers. Each industry buys goods and services from most other industries in the economy, often including its own output. Then each sector sells to all other sectors and final consumers.

When the equations describing these transactions are solved, it is possible to determine the total output from each sector of the economy required to produce a given vector of final demand. In the methodology used in this analysis, investment in solar-energy technologies was treated as final demand, since individual sectors in the I/O table were not created for each solar technology. Thus I/O made it possible to determine the total dollar value of output from each of the 200 sectors required to produce a given amount of each solar system. I/O analysis is described at greater length in Miernyck's introductory text (1965).

As noted in chapter 1 two kinds of environmental analyses were performed using SEAS: a stand-alone analysis and a scenario-based analysis. In the stand-alone analysis, the residuals associated with producing 10^{12} Btu from each solar-energy facility were compared with the residuals associated with producing 10^{12} Btu of energy from competing conventional energy facilities. In the scenario analysis, the residuals associated with entire economies (based to a greater or lesser degree on solar energy) were projected using the SEAS model.

The stand-alone analysis is a simple and intuitively appealing method to study the residuals associated with the solar technologies. However, it is methodologically inferior to the scenario analysis in several important respects. First, the stand-alone analysis was carried out using national data whereas the scenario analysis projected residuals at the regional level using regional and national data. A regional stand-alone analysis is possible but would require a series of regional I/O models. Second, in the stand-alone analysis the residuals, which result from the manufacture, construction, and operation of a solar system, are added together as though they occurred simultaneously, even though the construction period for a large centralized facility, such as a nuclear plant, may exceed ten years. This would be an acceptable methodology for analyzing a steady-state economy in which new energy technologies are not being introduced. It is less accurate for the analysis of an economy in which solar-energy technologies are being introduced rapidly. Finally, some types of residuals are simply not amenable to analysis without some regional specification. Area-wide or non-point-source water-pollution residuals are generally of this type and have been omitted from the stand-alone analysis though they are included in the scenario-based work.

The stand-alone analysis is not only inferior to the scenario-based analysis for projecting residuals from an economy that uses solar energy, it also is inadequate as a basis for computing the environmental damages that result from this economy. This inadequacy stems from the nonlinearity of damages as pollutant emissions rise within a given region. The damages associated with a given amount of pollutants emitted in a region depend on the amount of that pollutant already emitted by other sources in that region. Thus it is not possible to calculate the damages associated with pollution from any energy technology when this technology is torn from the context of the economy and region in which it operates.

On the other hand, the stand-alone analysis provides something that the scenario-based analysis is incapable of providing: a feeling for the steady-state environmental impacts of solar-energy technologies. Since the scenarios and the SEAS model extend only to the year 2000—at which time the growth rate of solar-energy technologies is likely to be substantial—the scenario-based analysis provides only a snapshot view of a transient phenomenon. To overcome this difficulty without the stand-alone analysis, it would be necessary to extend the model far into the future to a point where the growth rate of solar energy had declined to zero. This point might not be reached until after 2025, by which time the current model structure is no longer applicable. Together the stand-alone and scenario analyses provide estimates of the eventual steady-state emissions attributable to solar-energy technologies and emissions associated with an entire economy based, to a significant extent, on renewable sources of energy.

2.1.2 Model Choice: SEAS

Once it had been decided that the principal objective was to project both direct and indirect pollutant-emission consequences of high and low solar-energy scenarios and that an I/O approach was appropriate, there was very little choice of which of the available models was most applicable. The SEAS model was chosen because it was the only completely integrated energy-economic-environmental model available that incorporated an I/O model. To use a different national input-output model than the one SEAS incorporates (INFORUM) would have required the creation of another set of integrated data bases and calculations of energy investment, economics, and residuals. Moreover, no other emissions-generation model is demonstrably better than the SEAS modules. Finally, SEAS is the model used in the Department of Energy (DOE) by the assistant secretary for the Environment. Thus the results of a solar study using SEAS were likely to be comparable to other DOE energy-environment studies and would be widely reviewed within DOE.

Other computer models are applicable to certain aspects of the environmental analysis of solar energy. For example, the linear programming energy models available at Brookhaven National Laboratory would be an excellent vehicle for analyzing the tradeoffs between emissions and the costs of supplying energy, oil imports, and so forth (Marcuse 1978). However, to exercise this capability it is first necessary to have emissions projections, and SEAS was chosen for this purpose.

2.1.3 SEAS: Description

SEAS is composed of a series of independent computer modules. Booz-Allen and Peter House describe SEAS at greater length (Booz, Allen and Hamilton, 1975; House, 1977). A SEAS user's manual is currently being prepared by MITRE Corporation. In the interim, a series of internal memoranda are the only current documentation of the SEAS model. The interactions of these modules are shown in figure 1–1.

The Energy System Network Simulator (ESNS) module is a systematic method of portraying the flow of energy in the U.S. economy. ESNS begins with a list of end uses for energy in United States. Each of these end uses can be supplied in a number of different ways. For example, residential space heat can be supplied by gas or oil furnaces or electricity. Each of these sources of supply can come from any number of processes or technologies. For example, electricity for heating the home could come from nuclear-, gas-, oil-, coal-, wood-, or solar-fired steam generators. ESNS specifies the structure of energy use in the entire economy by tracing the flow of energy in detail from end-use demands back to sources of supply. Each time energy is transmitted from one place, or process, to another, some of it is degraded. ESNS specifies the first-law efficiencies associated with each transmission link in the economy.

ESNS is not a simulation model; it is merely an organized way of keeping track of sources and end uses of energy in the economy. ESNS is the entry point for scenarios in SEAS. By specifying the end uses, processes, and links in the network at a given point in time, a hypothetical structure of the energy system can be chosen. Given this structure and the magnitudes of end-use demands for energy, the model calculates the magnitudes of the various sources of supply required to meet the demands. End-use demands are specified for 1975, 1985, 1990, and 2000, with linear interpolation used between these years. The entire ESNS network is shown in figure 2–1. A diagram of the solar portion of the ESNS network, as reworked in this study, is presented in figure 2–2.

The energy investment module (EIM) of SEAS is a simple set of algebraic calculations that determine the energy investments required to meet the level of end-use energy demands supplied to ESNS. The EIM then breaks down these investment demands into demands for the output of each sector of the INFORUM I/O model.

To determine investment in a particular type of energy facility, given the time profile of demand for the output of that facility, EIM uses four variables: unit construction costs (1972 dollars—$millions/$10^{12}$ Btu/yr output) for each energy technology, the number of years it takes to construct the facility, the percentage phasing of costs over the construction period, and depreciation of existing facilities over time. The percentage of a given year's investment in each type of energy facility that is allocated to each INFORUM sector is read directly from a data base. For the nonsolar-energy sources, the Bechtel (1975) data based was used.

INFORUM, a 200-sector I/O model of the economy, is the central feature of the SEAS model. INFORUM forecasts the macroeconomic parameters of the economy, such as GNP, employment, consumption, investment, government purchases, and exports and imports as well as outputs for each of the 200 sectors consistent with the macroeconomic forecasts. Since INFORUM is a complex model, only a brief summary is offered here; a description is available in Almon et al. (1974).

Forecasts using INFORUM are made in three basic steps. First, certain structural relationships in the economy are specified. These relationships include I/O coefficients (the A matrix), the investment each industry group will need to spend on the products of each industry in order to supply a given level of final demand for its product in a specific year (the *B matrix*), and similar matrices for construction activities, government spending, and consumer demand. Exogenous forecasts of rates of productivity change in each sector complete the structural picture of the economy.

Given the structure of the economy, the second and third steps determine its size. In the second step, exogenous variables are projected, such as population, labor force, unemployment rate, number of households, interest rates, depreciation rates, and international exchange rates.

In the final step, the exogenously determined variables are combined with the detailed structure of the economy to yield self-consistent forecasts of the entire economy. Figure 2-3 is a schematic diagram of these calculations. In this diagram the quantities in boxes on the left side are exogenously specified for the base year. The values for subsequent years are calculated by the model. INFORUM has been modified somewhat for use in conjunction with other SEAS modules. Energy-related investment demands generated by the EIM are used in the SEAS version of INFORUM, replacing some of the econometrically forecasted investment demands originally used in INFORUM. The final result calculated by INFORUM is the dollar value of the output from each of the 200 sectors of the model required to meet the forecasted or specified levels of aggregate demand and energy end use.

Two additional features of the INFORUM model are of interest as used in SEAS. First, INFORUM A matrix coefficients are not constant but are forecasted, using logistic curves based on time series data or using special industry studies. Second, the A matrix coefficients for columns that repre-

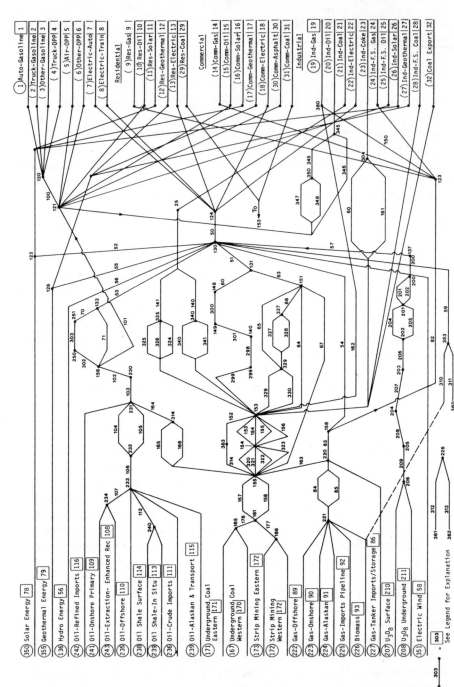

50 Electric Transmission
51 Electric Coal
52 Electric-Solar
53 Electric-Oil
54 Electric-Gas
55 Electric-Geothermal
56 Electric-Hydro
57 Electric-Nuclear
59 Electric-Biomass
60 Electric Coal Old-Other
63 Electric Coal-New
64 EU Fluidized Bed
65 EU Coal New-BACT
66 EU Coal New-NSPS
67 EU Combined Cycle
70 Electric Old Oil-Other
71 Electric Oil New
82 Gas Consumption-Direct
83 Gas Dist Pipeline
84 Gas Processing with H_2S Removal
85 Gas Processing without H_2S Removal
100 Service Stations
101 Oil-Direct Use
102 Oil Transport and Distribution
103 Crude and Petroleum Storage
104 Refining-Catalytic
105 Refining-Noncatalytic
106 Oil-Transportation
107 Oil-Onshore

112 Oil-Shale
140 Ind Coal-Old
141 Ind Coal-New
152 Pipeline
153 Coal Trans-Train Dest
154 Coal Trans-Barge Dest
155 Coal Trans-Truck Dest
156 Coal Trans-Other Dest
157 Coal Cleaned
158 Coal Not Cleaned
160 Beehive Coking
161 Slot Oven Coking
162 Low-Btu Gasification
163 High-Btu Gas
164 Liquifaction
165 H-Coal
166 Methanol from Coal
176 Underground Coal
177 Strip Mines Coal
200 Waste Management
201 No Recycling-Reprocessing
202 U&P Recycling Reprocessing
204 LWR
205 FBR
206 Fabrication
207 Enrichment
208 Conversion
209 Milling
298 Electric Coal Old-Fly Ash

299 Electric Coal Old-Sludge
300 Electric Coal Old-
301 Electric Coal Old-SO_x
302 Electric Old Oil-Pt.
303 Electric Old-SO_x
310 Electric Biomass-Wood
311 Electric Biomass-Municipal
312 Biomass Farms
313 Biomass-Residue Collection
314 Pipeline-Water Use
320 Coal Trans-Train Orig
321 Coal Trans-Barge Oil
322 Coal Trans-Truck Orig
323 Coal Trans-Other Orig
324 ICC New-NSPS
325 ICC New-SIP
326 ICC New-BACT
327 EU Coal-NSPS-Scrubber
328 EU Coal-NSPS-No Scrubber
329 EU New West
330 EU New East
340 ICC Old-Conv
341 ICC Old-Deprec
345 Ind Res Oil
346 Ind Dist Oil
347 Ind Res Oil-Old
348 Ind Res Oil-New
350 Ind Gas Pipeline

Figure 2–1. ESNS Network

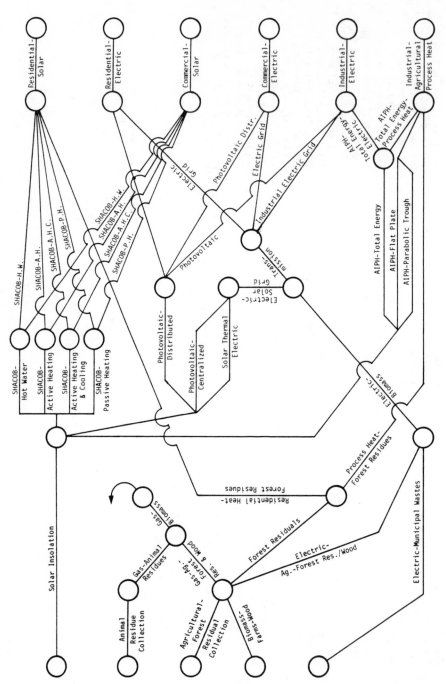

Figure 2-2. ESNS Solar Network

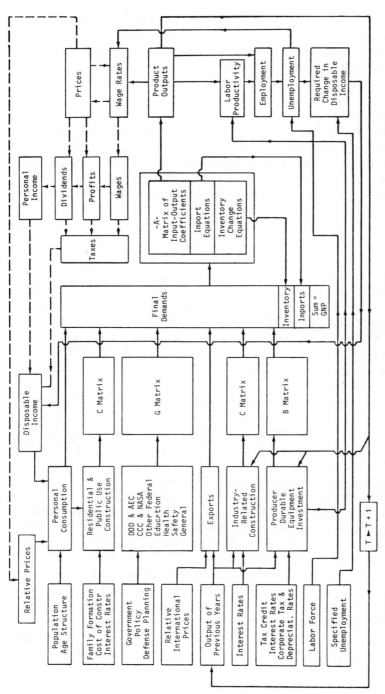

Figure 2-3. INFORUM Flow Chart

Source: Clopper Almon, et al., *1985: Interindustry Forecasts of the American Economy*, Lexington: Mass.; Lexington Books, D.C. Heath & Co., Copyright 1974, D.C. Heath & Co. p. 10. Reprinted with permission.

sent energy sectors (such as electric utilities) are adjusted automatically to make them consistent with the coefficients that would be implied by the Energy System Network Simulator. For example, if the fraction of electricity generated by solar sources increases over time, purchases of other fuels by electric utilities will decrease and this will be reflected in the column vector for electric utilities.

While the 200-sector level is fairly detailed for examining an entire economy, it is not sufficiently detailed for projecting pollutant emissions. Moreover, pollutant emissions are related more closely to physical flows than to dollar flows. INSIDE, another SEAS module, disaggregates INFORUM outputs further for environmentally important sectors and converts these outputs from dollars to tons. Disaggregation is performed for INFORUM sectors when a sector has a number of different products associated with it and the residuals produced by these products differ from one another, or when a particular product from a sector is produced by more than one process, and the residuals associated with those processes differ from one another. In general, the sector disaggregation in INSIDE remains the same from year to year. Results of detailed studies of technological change, where available, were incorporated by using sector disaggregation coefficients that change through time. INSIDE disaggregates 50 sectors into about 350 subsectors, giving the model a total of about 500 sectors.

Once the INFORUM sectors are disaggregated by INSIDE, they are regionalized by another SEAS module called REGION. Regional forecasts of economic activity are important in projecting regional pollutant emissions. Since the damages associated with these emissions are strongly dependent on the regional distribution of national emissions, this is an important step. In general, REGION forecasts at the county level. County level forecasts can be aggregated to larger regions, such as air quality control regions (AQCRs) and states.

The REGION module begins with base-year data on the regional shares of national output for each sector. Then, using forecasted growth rates for each sector in each region, shares in future years are forecasted. Base-year shares are generally based on employment data for the relevant industries at the county level. When employment data are unavailable for industries, other data are substituted in the share-creation process. Data on growth rates of individual sectors or subsectors are computed from the data available in OBERS (U.S., Department of Commerce, 1974). These data originally are at the two-digit standard industrial classification (SIC) level and are based on dollar earnings. This requires that all INFORUM sectors contained within a two-digit SIC code are constrained to grow at identical rates within a given region. Sectoral shares for each region are based on the fraction of the national earnings contributed by a region.

Having determined the outputs of all INFORUM sectors and INSIDE subsectors at the regional level, the next SEAS module to be run, called RESGEN, computes the gross and net amounts of pollutant emissions associated with each of these outputs. Gross emissions are the amount that would be produced in the absence of an abatement technology; net emissions are those produced after pollution abatement.

RESGEN contains a large data bank of residual coefficients that give total residuals when multiplied by the appropriate output. Three types of residual coefficients are used. For energy-related emissions, coefficients are given in tons/10^{12} Btu. Total emissions are determined by multiplying quantities specified in ESNS by coefficients. For INFORUM sector outputs, coefficients are given in tons per dollar of output and multiplied by the dollar value of output for each INFORUM sector. For INSIDE subsectors, coefficients are given in tons per ton of subsector output and multiplied by the number of tons of output.

RESGEN also can be fed two other modules, LANDUSE and TRANS-PORTATION. LANDUSE computes non-point-source water pollutants, and TRANSPORTATION computes transportation-based air-pollutant emissions. LANDUSE was not used in this analysis and is not discussed further.

Two additional modules complete the basic SEAS structure. These are regionalization routines for the electric utilities and industrial combustion energy sectors. Since the analysis for this report made special use of these programs, they are described in greater detail at the end of section 2.1.4.5.

2.1.4 SEAS: Application to Solar-Energy Scenarios

2.1.4.1 Changes to ESNS. The application of SEAS to determine the environmental impacts of solar-energy scenarios began with a reorganization of the solar portion of ESNS. The following discussion of the reworked solar portion of ESNS refers to figure 2–2.

Solar insolation is used for SHACOB, AIPH, and electric applications. SHACOB systems are either active or passive and used in the residential and commercial sectors. AIPH systems are either flat-plate, parabolic-trough, or parabolic-dish systems. The parabolic-dish system is used to produce both electricity and heat and is called a *total-energy system*. Heat from all three systems is used in the agricultural and industrial sectors. Electricity from the total-energy system is used on site and is not transmitted via the electricity grid. Solar electricity is generated either by photovoltaic or central-station solar-thermal systems. Photovoltaic systems are both centralized and distributed. The centralized system transmits its electricity to end users via the electricity grid; the decentralized system is used on site by the residential, commercial, and agricultural and industrial sectors. Solar-thermal generation of electricity is centralized and distributed via the electricity grid. WECS transmit electricity via the grid.

Biomass used for energy comes from three sources: animal residues, agricultural and forest residues, and silvicultural farms. Animal residues are gasified, using an anaerobic digestion system. Gas from this system is transmitted via the gas-pipeline transmission network. Agricultural residues are gasified using a pyrolysis process and then transmitted via the pipeline network. Forest residues are used in several applications. One portion is used on-site at pulp and paper mills as process heat. Another portion is used in wood stoves at residences. The remainder is gasified by pyrolysis and transmitted via the gas-pipeline network. Finally, wood from silvicultural farms is used to fire a steam generating plant that transmits its electricity via the electricity grid. Descriptions of each of these systems can be found in section 3.1.

2.1.4.2 Scenario Generation Procedures. The second step in this analysis was the specification of future energy scenarios emphasizing solar energy. A number of sources existed from which to derive energy scenarios. A partial list is shown in table 2-1. The year 2000 baseline estimates of total U.S. energy supply ranges from 108 to 177 quads in these sources.

The possible scenarios and their priorities are illustrated in figure 2-3, using the solar fraction of the total energy supply as a parameter versus total national energy use. The most important cases are: (1) the base case, (2) high solar use, and (3) very high solar use. Cases 4, 5, and 6, as shown in figure 2-4, were assigned a lower level of importance; cases 7, 8, 9, and 10 were assigned the lowest priority. All of these cases could be repeated at varying levels of environmental controls. Due to resource limitations, only scenario types 1, 2, and 3 were analyzed in this book, each at two levels of environmental controls.

Rather than attempt to repeat the scenario-creation process independently and derive yet another estimate of the energy future, the approach used to establish energy scenarios for this study was to take advantage of the president's ongoing Domestic Policy Review (DPR) of solar energy. This major and broadly based federal-level activity established a series of energy projections of conventional and solar-energy use as a basis for a major review of solar-energy policy. The development of these scenarios involved thirty federal agencies with contractor support. In making judgments on scenario parameters, the DPR used studies from the following: U.S., Department of Commerce (1977), U.S., Department of Energy (Energy Information Administration, 1978), MITRE (MITRE Corp., 1977), and Stanford Research Institute (SRI, 1977). Also used were MOPPS (U.S., Energy, Research and Development Administration, 1978),

Table 2–1
Major Recent National Energy Scenarios

Source and Date	Scenario Name	Total Quads by 2000/2020	Total Solar Quads by 2000/2020
MITRE/Spurr, March 1978	NEP[c]	115/189	6/33
	RTS[d]	113/188	5/23
Solar Working Group, February 1978 (SRI)	Base case	115+4/138+10	8.1–8.8/19.1–19.5
	Low solar price	[a]	11.1/28.9
	High electrification and high demand	[a]	11.8/35.2
	High nonsolar price	[a]	18.6/50.2
SRI, March 1977	BAU[e]	145/198	6/11
(Solar Energy in America's	Solar emphasis	148/204	15/44
Future)	Low demand	89/102	7/20
Dartmouth, June 1977	BAU	143/[b]	3/—
(Fossil I)	NEP	118/[b]	3/—
	NEP and accelerated supply	119/[b]	
	NEP and accelerated coal	122/[b]	
CONAES	High solar	146/190+	13.1/28.8
	Low solar	146/190+	0.1/0.6
Lovins, 1977		108/76	40/71
IERPS, 1977		118/148	14/29
MOPPS, 1977		118.4/[a]	2/—
ERDA 49, 1977		150/180	10/45
NF/NASA, 1972		177/300	11/135

[a] Not available in published documents.
[b] Study did not go beyond the year 2000.
[c] NEP = National Energy Plan.
[d] RTS = Recent trends.
[e] BAU = Business as usual.

CONAES (National Academy of Sciences, 1977), and IERPS (U.S., Energy Research and Development Administration, 1977e). In addition, unpublished analyses using the MITRE/SPURR model were available for consideration. These scenarios are listed in table 2–1.

Five energy scenarios were defined by DPR:

DPR 1—base case at the expected price and availability of conventional energy;

DPR 2—base case at high price and low availability of conventional energy;

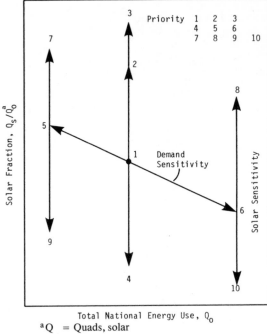

aQ = Quads, solar
Q = Quads, total

Figure 2–4. Energy Scenario Strategy

DPR 3—base case at low price and high availability of conventional energy;

DPR 4—maximum practical case for solar-energy penetration; and

DPR 5—technical limits case for solar-energy penetration.

In this book, DPR 1 was chosen as the Base Case, and DPR 4 and 5 were chosen for the analysis of maximum-practical and maximum-feasible solar-energy market penetration. These scenarios were labeled SERI 1, 2, and 3, respectively. All three scenarios assume total energy demand in the year 2000 to be 114 quads. (This was revised to 95 quads in the 20 June 1979 presidential DPR message.) The macroeconomic assumptions associated with these scenarios are shown in table 2–2. Data on each scenario, and a general discussion of the methods the DPR used to develop the data are presented in section 3.2.

High Solar-Energy-Use Scenarios. The high solar-energy-use scenarios analyzed in this study were generated by the DPR based on the mid-price

Table 2-2
Macroeconomic Assumptions for the Base-Case Scenarios

	1977		2000	
Assumption	$15/bbl[a]	(DPR 3) $18/bbl	(DPR 1, SERI) $25/bbl	(DPR 2) $32/bbl
Total energy demand, in quads (percent annual change)	75.9	132 (2.4%)	114 (1.8%)	95 (0.9%)
GNP, in trillions of 1977 dollars (percent annual change)	1.89	3.99 (3.3%)	3.81 (3.1%)	3.65 (2.9%)
Total energy/GNP, in thousands of BTU per dollar	40.1	33.1	29.9	26.0

[a]1978.

base case. The maximum-practical case (SERI 2) would result from a determined effort at federal, state, and local levels to introduce solar technology
into society. The total solar contribution in the maximum-practical case is
16.8 quads or 15 percent of the energy supply by 2000. The fifth scenario
generated by the DPR panel is the maximum feasible or *technical limits* to
solar-energy-use case (SERI 3). This case would result from all-out mobilization to increase solar-energy use, and the nation's physical production
and installation capacity would be the major limits on solar-energy use.
Market economics or other financial and institutional barriers would not
constrain solar-energy use in any manner in the maximum-feasible case.
Total solar-energy use is 27.5 quads (24 percent) in 2000. The solar-energy
contributions in the maximum-practical and maximum-feasible scenarios
are summarized in table 3-5.

Once the solar-energy contributions for the high solar-energy scenarios
were projected, national consumption of conventional fuels was estimated
by the DPR. The general approach assumed that high solar-energy use will
be caused primarily by a broadly based political decision to make a transition to solar energy. The social motivation for increased solar-energy market penetration was assumed to be both a reaction to the adverse characteristics of conventional energy systems and a positive reaction to some of the
characteristics of solar energy.

A crude judgment was made by the DPR Impacts Panel that oil, coal,
and nuclear energy are more similar in the magnitude of social disruption
and difficulty than they are different and that gas causes fewer serious
social problems. In the year 2000, solar energy was assumed to substitute
for coal, oil, nuclear energy, and gas in the following proportions: 0.3, 0.3,
0.3, 0.1, respectively. (More details concerning the assumptions and procedures used in generating the energy scenarios are found in section 3.2.)

Environmental Control Assumptions. Each of the scenarios described already was run with two sets of environmental controls, dubbed *strict* and *lax.* This was done in order to ascertain whether changes in the mix of energy technologies or environmental controls would have larger environmental impacts and whether changes in environmental controls would affect the relative environmental impacts of high and low solar-energy scenarios.

In the strict scenarios, compliance with all federal and state environmental regulations now on the books for both current and future years was assumed. For example, if a state implementation plan (SIP) called for a higher level of control (lower emission) than the federal New Source Performance Standards for sulfur-dioxide emissions, the SIP was assumed to govern actual emissions.

In the lax environmental scenarios, it was assumed that controls for air pollution would not become any stricter than those required by the Environmental Protection Agency as of January 1977. Thus the Clean Air Act revisions (as amended in 1977) were ignored. This means that no revised New Source Performance Standards were assumed for either coal-fired utilities or industrial boilers. (Both are assumed to comply with current New Source Performance Standards.) Emission factors for automobiles and trucks were frozen in 1978. For water pollution, it was assumed that all facilities covered by the Clean Water Act amendments of 1972 would meet Best Practicable Technology regulations. Best Available Technology regulations were assumed not to take effect prior to the year 2000.

2.1.4.3 Energy Investment Module Modifications. Much of the work for this application of SEAS involved gathering and manipulating data for the EIM. The data utilized are described in chapter 3. No modifications were made to the EIM software. This analysis used the EIM in a novel way, however, by specifying capital costs for the solar technologies that vary with time and whose sectoral components vary with time as technical progress either modifies or changes solar-energy systems entirely.

2.1.4.4 Energy and GNP Targeting. To run INFORUM, two special procedures are needed, called *energy targeting* and *GNP targeting.* The energy-targeting procedure assures that the total and sectoral energy demands specified in a particular scenario and entered into the system via the ESNS module are consistent with the energy use that is implied by the size and structure of the economy. Energy targeting involves running INFORUM iteratively and adjusting energy-related A matrix coefficients each time until the energy use implied by INFORUM for each energy source approximates that specified in the scenario being analyzed. To do this, the energy-targeting procedure relies on the SEAS energy module, which computes the demand for each energy source implicit in INFORUM sector outputs. The energy-targeting procedure is described at greater length in Metzger's INFORUM Targeting Procedure (1977). Since this procedure involves adjusting the hypothesized structure of the economy to conform to prespecified energy supply and demand scenarios, it places a premium on the careful specification of these scenarios.

GNP targeting assures that GNP, unemployment, and productivity are mutually consistent. This procedure begins by using a series of econometrically based equations to forecast productivity in each INFORUM sector. INFORUM can be run to produce a target level of GNP, unemployment, or productivity. In this analysis, it was desired to examine the environmental effects of increasing solar-energy use, independent of the effects of solar-energy investment on GNP and employment. Therefore target levels of GNP and employment were picked, and productivities projected by INFORUM's productivity equations were scaled proportionately so that INFORUM projected the target values of GNP and employment. The implications of this procedure are discussed in chapter 4. GNP and employment targeting are discussed at greater length in Metzger (1977).

2.1.4.5 Regionalization of Energy Use. In determining the regional environmental impacts of each scenario considered, it was necessary to allocate the national energy supply on a federal-region basis. To determine the economic damages associated with each scenario using the methodology described in section 2.3, it was necessary to project environmental impacts at the AQCR or aggregated subarea (ASA) level. To make this projection using SEAS, it was first necessary to make environmental projections at the county level and then to aggregate counties to AQCRs or ASAs. Thus a county-level regional allocation of each energy source used in each year in each scenario was required even though the residuals analysis presented here is only at the federal-region level. The description presented here is a summary of the methodologies used to make this county-level disaggregation. Although the location of energy facilities is important, it is only responsible for the regional distribution of the direct environmental impacts of these facilities. The indirect impacts are distributed as the production of the inputs to these facilities is distributed.

To determine the state share of national solar-energy use for each solar technology, the following methodology was employed. First, the census-region share of solar-energy supply in each year was obtained from the SPURR model. (SPURR model results were felt to be as good as those from any other solar market penetration model and were easily available.) These shares for the year 2000 are shown in table 2-3. For solar-thermal electric, wind electric, biomass electric, and photovoltaic energy supply, state shares of census regions were scaled on the basis of projected new electric demand

Table 2–3
Census Region Solar Demands for Year 2000
(quads)

Solar Application	Source	Total	Northeast	Middle Atlantic	South Atlantic	East North Central	East South Central	West North Central	West South Central	Mountain	Pacific
Buildings	Thermal	1.6	.06	.32	.08	.38	.08	.14	.21	.11	.2
	Passive	1.1	.04	.22	.06	.26	.06	.10	.14	.08	.1
	PV										
	Wind										
Industry	Thermal	1.23	.03	.14	.16	.21	.07	.07	.33	.05	.14
	Biomass										
	PV										
	Wind										
Electric utility	Thermal	.92	—	—	.23	.02	.02	.13	.19	.11	.22
	Biomass	.04	0	.003	.022	.003	.011	0	0	0	0
	PV	.31	—	—	.09	—	—	—	.07	.02	.06
	Wind	1.74	.13	.32	.31	.35	.28	.05	.13	.04	.13
	OTEC	—	0	0	.006	0	—	0	0	0	0
Synfuels	Biomass	4.5	.046	.036	.086	.054	.078	.18	.036	0	.019
Total		11.44	.31	1.04	1.04	1.28	.60	.67	1.11	.41	.87

Source: MITRE Corporation. 1978. "Estimates of Solar Savings." Scenario: Base in MITRE memorandum on "Documentation of Baseline Scenarios for the DPR." Memorandum no. W52-M–269, 25 July 1978. Reprinted with permission.

Note: Estimates are based on the assumptions of the SERI study.

in each state. Without more detailed information on climate and individual electricity markets, scaling on the basis of new electric demand seemed the most reasonable assumption. These projections were obtained using a three-step procedure. In the first step, federal-region-level projections of new electric demand were obtained from the Energy Information Administration (EIA) Scenario Series C (Energy Information Administration 1978). In the second step, these demands were scaled on the basis of National Electric Reliability Council projections at the regional level (Stanford Research Institute 1976). Finally, new electric demands were scaled to the state level on the basis of population data obtained from OBERS (U.S., Department of Commerce 1974).

For SHACOB and AIPH, state shares of energy supply were scaled on the basis of projected population, obtained from the Department of Commerce (1974). This is a crude assumption but a reasonable one to make at this level of aggregation. AIPH state shares were obtained by scaling federal-region totals to OBER's forecasts of future economic activity (1974). This was done because it was felt that industrial process-heat applications were related more to future economic activity in a state than to population. For the supply of animal, forest, and agricultural residues used as a feedstock for gas production, state shares were based on data provided by Stanford Research Institute (1976). For the supply of wood from silvicultural farms, state shares were based on information from the MITRE Corp. (1977). For the generation of gas from all sources of biomass, state shares were based on an average of the state shares of feedstock production. For forest residues used as process heat, state shares were based on shares of forest residue production. For residential use of forest residues, state shares were based on state shares of national forest residue production. In this calculation, the southern states were excluded from the national total and state allocations because only marginal use of wood for residential heating occurs in southern states. These regional shares are the same for all the scenarios.

Given the state shares of national solar-energy use, the following methods were used to distribute energy use to counties. For SHACOB, AIPH, photovoltaics, biomass electric, biomass process heat, and wood stoves, the county share of state population was used to proportion energy use. For animal residue and agricultural and forest residue collection, silvicultural farms, and anaerobic digestion, county shares were set equal to the percentage of a state's agricultural farm acreage. Population and agricultural farm acreage were obtained from Lerner (1972).

None of these county allocation procedures is critical since the procedures are never used directly. Residual analysis is reported only at the federal-region level. Even in the economic benefit-cost analysis, county-level results are only used to reaggregate to air quality control regions or aggregated subareas.

Different methodologies were used to allocate conventional sources of energy to individual regions in the electric utilities, industrial, and residential and commercial end-use sectors. For the electric utilities, the methodology began with the solar-electric demands derived as described above. Next, national electric demand was allocated to federal regions on the basis of the Energy Information Administration (EIA) Series C projections (1978). These projections were based on the Project Independence Evaluation System (PIES model). PIES projections were used because PIES is one of the most comprehensive and well-structured models of energy supply and demand ever built in the United States. These allocations were identical for all scenarios and assumed to be met. Regional nonsolar-electric demands that were required to be met were calculated. To allocate the conventional electricity sources to meet these control totals, the SEAS electric-utilities module was run. This module was run first to determine the fuel mix at the federal-region level, and then again to determine each state's fuel mix, given the derived federal-region fuel mix. Finally, a separate subroutine, based on the location of existing plants, allocated each type of electric plant to the county level.

The electric utilities module works as follows. First, the amount of existing electric capacity of each type in each region is obtained, along with its projected lifetime and capacity factor. By subtracting the amount of electricity that can be supplied by each type of plant in each region in each year from the new electric demand, the total required additional capacity in each year is calculated. To determine which types of fuel will supply this demand in each region, a linear program is used since there are an infinite number of ways to satisfy the region control total and the fuel-use control totals simultaneously. Basically the linear program is run to minimize the difference between a trial solution and any other possible solution. The solution that is obtained depends critically on the trial solution used in the algorithm. For the scenarios analyzed here, the regionalization used in the PIES model was used as a trial solution.

Once the state allocations of electric-generation plants are obtained, they are allocated to counties on the basis of a data file of plant locations already announced but not yet built. When all the existing sites of a given type in a region have been used, another data file is used that ranks each county in the United States on the basis of its suitability for each type of electric-generating facility.

The regionalization procedure for combustion of industrial fuels began with state control totals taken from the EIA projections. Since the EIA national totals for each fuel were close but not identical to those in the SERI scenarios, these totals had to be scaled so that each fuel use equaled the national total when it was summed over all states. This procedure was used in 1975 and 1985. By 1990, however, the national control totals for each type of fuel use differ markedly in the EIA and SERI scenarios. Thus simple scaling of the SERI scenarios to EIA regions was inappropriate. To circumvent this problem, EIA regional total conventional fuel use was multiplied by the ratio of SERI base case national industrial conventional fuel use to EIA national industrial fuel use to get a new federal-region energy demand. Some judgment was exercised to determine which fuels supplied that regional energy demand. For natural gas, totals in the EIA and SERI base cases were close, and simple scaling was used. For coal, the difference between the EIA and SERI base cases was allocated to regions based on where increased coal use occurred in the previous SEAS run of the National Energy Plan, which had a lot of coal use in it. Residual and distillate oil made up the difference between regional energy demand and the sum of coal and natural gas use. The split between residual and distillate oil was based on the EIA scenario. Once these procedures were applied to get federal-region totals, state shares of federal-region totals were based on the EIA scenario. Finally, to get county-level fuel splits in the industrial sector, the SEAS Industrial Boiler Module was run. It is primarily an organized way of manipulating existing data rather than a simulation routine.

The EIA scenario also was relied on for the regionalization of residential and commercial fuel use. In 1975 and 1985, the EIA and SERI base cases were almost identical, so state shares were taken directly from the EIA scenario. This was true also for the 1990 commercial fuel use. When comparing the SERI scenarios with the EIA scenario, residential solar energy substitutes for oil. In the base case for 1990, all state oil-consumption values from EIA were reduced by the same fraction so that residential oil use summed to the base-case national total. The base-case state oil totals were used, then, as a basis for computing the maximum-practical and maximum-feasible state oil uses. This was done by subtracting the total amount of solar residential energy use in each state (previously obtained) from the base-case oil-use figures for each state. The remaining oil use then was renormalized so that the national oil use was obtained. This same procedure was used in the base case for the year 2000. In the maximum-practical and maximum-feasible cases, however, the state natural-gas shares, as well as the oil shares, differ from the EIA scenario since solar energy substitutes for oil and natural gas. The state shares used for each type of fuel depend on the amount of that fuel already used in the state.

2.1.4.6 Additions to and Use of SEAS Environmental Models. Since the goals of this study were to project the pollutant emissions and environmental damages associated with high and low solar-energy scenarios, much attention was given to the use of SEAS environmental models. Most of this work involved replacing existing RESGEN solar coefficients or adding new ones. Table 2–4 describes the RESGEN solar data.

Table 2-4
RESGEN Solar Data Base

Technology	P	SO_x	NO_x	HC	CO_2	CO	Cl_2	COD	SS	DS	Non-combustible Solid Waste	Pesti-cides	Alde-hydes	Land Use	Water Use	In-juries	Work-Days Lost	Opera-tional Labor
SHACOB—passive								X								X	X	X
SHACOB—active														X	X	X	X	X
AIPH—flat plate								X						X	X	X	X	X
AIPH—parabolic trough								X						X	X	X	X	X
AIPH—parabolic dish								X						X	X	X	X	X
Solar thermal							X							X	X	X	X	X
PV—centralized														X	X	X	X	X
PV—decentralized															X	X	X	X
Wind														X		X	X	X
Manure collection																		
Agricultural forest residues	X	X	X	X	X									X	X	X	X	X
Biomass farms	X	X	X	X	X	X			X	X			X	X	X	X	X	X
Anaerobic digestion		X	X		X	X				X	X	X		X		X	X	X
Pyrolysis	X	X	X	X	X						X			X	X	X	X	X
Wood-fired steam electric	X	X	X	X	X						X			X	X	X	X	X
Wood stoves	X										X					X	X	X

Note: X indicated an entry in the data base.

Direct operating residual coefficients were developed for the sixteen solar technologies studied. The residuals developed for active SHACOB and the three AIPH systems include land use, chemical oxygen demand (caused by emissions of antifreeze to the environment via system leaks), and labor-related residuals. Labor-related residuals include injuries, workdays lost, and operational labor. The only WECS residuals are land use and labor related, while solar-thermal residuals include water use, chlorine emissions to the environment via cooling water, and land use. For the PV central power station, both land and water use are included. The only distributed PV plant residual is water use.

Residual coefficients for the collection of agricultural and forest residues include coefficients for particulates, sulfur oxides, nitrogen oxides, hydrocarbons, and carbon monoxide. Biomass farm residuals include particulates, nitrogen oxides, hydrocarbons, carbon monoxide, carbon dioxide, aldehydes, dissolved and suspended solids, water use, and land use. Anaerobic digestion of manure residuals include sulfur oxides, nitrogen oxides, carbon dioxide, solid wastes, land use, and water use. Pyrolysis of agricultural wastes residuals are sulfur oxides, nitrogen oxides, hydrocarbons, carbon monoxide, solid wastes, pesticides, water use, and land use. Steam-electric-plant-residuals categories are identical to those for pyrolysis, with the exception that no pesticides are reported. For the five biomass technologies already listed, labor-related residuals were reported also. However, no labor-related residuals were provided for wood stoves and collection of manure. Wood-stove residuals are particulates and solid wastes (ash). No residuals were reported for collection of manure.

To obtain accurate estimates of pollutant emissions, especially for those emitted to water, several non-point-source computer modules were run. These include an urban runoff model, a rural runoff model, and an agricultural and mining runoff model, all described in Ridker and Watson (1978).

2.1.5 SEAS: Application to Individual Technologies

As noted in section 2.1, a simplified application of SEAS was made to determine the indirect national-residuals impact of each solar and conventional energy facility. In this application, only the INFORUM module and point-source portions of RESGEN were utilized.

This analysis began with the solar and conventional capital-cost vectors described in section 2.1. These cost vectors apply only to the solar technologies and do not include any costs for conventional backup systems that may be required. These vectors then were multiplied by the Leontief inverse matrix, $(I - A)^{-1}$, and a vector of point-source pollution coefficients per dollar of output of INFORUM sectors. This calculation was performed using the A matrix and net residual coefficients for 1975, 1985, and 2000. Consequently, the indirect residuals from solar construction vary as the structure of the economy and the level of environmental controls change. The base-case A matrix and high control net residuals are used for all calculations. A similar, but separate, calculation is made to determine the industrial combustion, commercial, and transportation energy-use-related pollution impacts because the RESGEN model uses pollution coefficients based directly on energy use for these sectors rather than coefficients based on INFORUM sector outputs. In this calculation, the Leontief inverse is multiplied by the capital-cost vectors to obtain the indirect INFORUM sectoral impacts associated with each solar system. The resulting vector is used as input to the ENERGY module, which projects fuel use by type resulting from any given vector of INFORUM sector outputs. RESGEN coefficients are multiplied then by the calculated fuel uses. Finally, total indirect pollution impacts are calculated by adding together those impacts from the industrial combustion, commercial, and rest-of-the-economy sectors for each pollutant.

The preceding analysis allowed examination of the indirect pollution impacts of each solar and conventional energy facility at the national level. It provided results that are similar in character, more comprehensive, and less detailed than those that would be obtained by a process analysis of the pollution associated with each energy facility.

This procedure allowed the pollution impacts of each solar-energy facility (for example, a solar-thermal power plant) to be compared to those of a conventional facility (for example, a coal-fired steam-electric plant). It did not allow the environmental impacts of a given amount of energy delivered to the end user by solar or conventional means to be compared. Although the indirect emissions from the production of each facility modeled were taken into account, not every facility utilized in the fuel cycle for each technology considered was analyzed.

2.2 Method for Projecting Residuals Concentrations

The Watson AMBIENT module was used in this project to translate residuals produced by energy scenarios into concentrations. The following description is adapted from Watson (Ridker and Watson 1978). The AMBIENT module accepts net residuals by region and pollutant, summed over all sources as an input from the SEAS/RESGEN system. The projected net residuals are transformed next into forecast ambient concentration levels by the application of a simple model. Base-year concentrations of each pollutant in each region are linearly scaled up by the ratio of forecast emissions to base-year emissions and multiplied then by *source-receptor transfer coefficients,* which vary by pollutant and economic sector but are identical for all regions.

The source-receptor transfer coefficient is the product of (1) the percentage contribution by a given pollutant source to exposure from that pollutant from all sources in a given region in the base year, and (2) a factor measuring relative frequency of stagnant air conditions and relative nearness of receptor to source. This coefficient, therefore, incorporates the meteorology (for air) or transport processes (for water) at the time of the base-year measurement.

Water quality is measured using the PDI index. This index is a crude but, in principle, comprehensive measure of ambient water quality. (PDI is defined as P $\times$ D $\times$ I where P is the number of stream miles within a region not in compliance with federal and state water-quality standards. D is the number of quarter-year periods in which this violation occurs and I is a judgmental parameter reflecting the overall effects of pollution severity. Further discussion of the PDI index is available in Truett et al. (1975).) It is intended to reflect the impingement of water pollution on human activity as well as the degree of pollution concentration. Higher values indicate lower water quality. It is used to assess environmental quality (instead of pollution concentrations) because water pollution concentration data are fragmentary and of suspect quality and dispersion modeling of water residuals is more difficult than in the case of air. Also, assessment of the relationship between water quality and impacts on human activities—whether or not they are based on the PDI index or some other procedure—almost always relies on subjective judgments of the type required in using the PDI index (Lake et al. 1976). The source of the base-year PDI values was Lake et al. (1976). Data on PDI values were available for 81 of 101 aggregated subareas.

2.3 Method for Projecting Economic Damages

The Watson BENEFITS module was used to estimate and project air and water-pollution damages and control costs (1978). For air pollution, calculations were performed for each of the 243 AQCRs and then summed to obtain national estimates. For water pollution, damages and pollution control costs were calculated for each of 101 ASAs and then summed to obtain national estimates.

Air-pollution damages and control costs were estimated separately for five air pollutants (particulate matter, sulfur oxides, nitrogen oxides, hydrocarbons, and carbon monoxide) and for four major sources of emissions (electric utilities, industry, residential and commercial sources, and transportation). Water-pollution damage and control costs were calculated for eleven pollutants (biological oxygen demand, chemical oxygen demand, suspended solids, dissolved solids, nutrients, acids, bases, oil, grease, heavy metals, and pesticides) and then allocated to seven major sources (electric utilities, industrial point sources, municipal sewage treatment, urban runoff, agricultural runoff, mining, and other non-point sources).

The BENEFITS module started with estimated national water- and air-pollution damages in 1971. It used the Gianessi et al. (1977) national-gross-air-pollution-damages estimate of $20.2 billion (1971 dollars) as an input. This estimate reflects the air-pollution national-damage value in 1971 dollars of the total 1968 emissions of the five principal types of air pollutants: TSP, SO_x, HC, CO, and NO_x.

The national-air-pollution-damage estimate of $20.2 billion used in the Watson system understates total damages because it does not reflect all of the potential damages. Among the most important damages missing are the psychic costs of illness and death and the damages due to environmental health effects not yet understood.

In projecting the economic damages from water pollution the Watson BENEFITS module began with estimated national damages of $11.1 billion. This includes $10.1 billion calculated by Heintz, Jr. et al. (1976) and $1 billion added by Watson (Ridker and Watson 1978) for health care and foregone income as a result of increased cancer incidence.

During the planning for the research for this book, it was hoped that the Gianessi-Heinz national-damage estimates from air and water pollution could be substantially reevaluated and presumably enlarged, taking into account any available additional research. It was hoped also that improved estimates of the damage functions for each pollutant could be made. These tasks proved to require more effort than could be allocated to them under the available research budget. A review of the literature on air- and water-pollution-damage estimation was prepared but no further work was done (Krawiec 1979). The literature review, combined with more recent efforts by other investigators, should allow further work in this area to proceed (Mendelson and Orcutt 1979; Fisher 1979).

The BENEFITS module applied 1971 national damage estimates for the five major air pollutants and all water pollution to estimate the corresponding per-capita regional-damage function having the following properties:

1. Regional per-capita damages are a function of regional pollution exposures to values-at-risk.
2. The sum of regional total damages in 1971 equals the exogenous national damages supplied to the BENEFITS module for 1971.
3. Each regional damage function has the same relative rates of increase or slopes as exhibited by a few existing specialized empirical damage functions.

Thus if damages per capita quadruple along the empirical function as exposures go from their 1971 median level to twice that level, then each regional damage function also quadruples damages per capita as exposures go from their 1971 median to twice the median, and so on. The damage functions, relative to the median, are shown in table 2-5. Air-damage functions were derived from data in Ahern (1974) and Watson's judgment (Ridker and Watson 1978); water-damage functions were based on Dornbusch (1973) and Watson's judgment (Ridker and Watson 1978). It is essential for the reader to understand that these damage functions were not built from the bottom up but instead were designed to rationalize the estimates of national pollution damage made by previous investigators. Formal derivation of damage functions with this property is shown in appendix B.

Implicit in the estimates of national pollution damages is an economic value of $30,000 (1972) on one life. This unrealistically low figure was used also in Watson's damage functions to assure consistency with the national damage control totals. Because value-of-life estimates in the literature are substantially in excess of $30,000 (Fisher 1979), several alternative values were used in the estimating procedure and are reported in section 4.1.4.

These alternative values are used to increase the health-related portion of national air-pollution damage costs. Water-pollution damage costs are unaffected by changes in the assumed value of life because nearly all water pollution damages are recreation rather than health related.

To forecast damages and pollution control costs, the BENEFITS model used gross and net residuals as generated by the RESGEN model. Gross residuals assume no pollution control; net residuals are residuals cor-

Table 2-5
BENEFITS Damage Functions

Exposure Level Relative to Median	Partic- ulates	Sulfur Oxides	Nitrogen Oxides	Hydro- carbons	Carbon Monoxide	Water Pollution	PDI Linear
½	0.2	0.2	0.2	0.2	0.2	0.4	0.5
1	1.0	1.0	1.0	1.0	1.0	1.0	1.0
2	3.0	3.0	2.8	2.8	2.8	2.2	2.0
3	5.5	5.5	4.8	4.8	4.8	3.5	3.0
4	9.0	10.0	7.3	7.3	7.3	5.0	4.0
≥ 5	14.0	16.0	10.3	10.3	10.3	6.7	5.0

Source: *R.G. Ridker and W.D. Watson, To Choose a Future: Resource and Environmental Problems of the U.S., A Long-Term Global Outlook,* (Washington, D.C.: Resources for the Future, Inc., 1978), Appendix A.
Note: Table shows concavity relative to region with median damages in the base year.

responding to specified control levels. Net residuals from both point and non-point sources were converted to effective concentrations or exposures and then mapped into the regional per-capita damage functions. The resulting quantities were multiplied by regional population and a regional income-environment elasticity multiplier to obtain regional damages for each scenario. These damages were added to pollution control costs to obtain total damages.

The income-environment elasticities in this book are based in part on Harris et al. (1968) and Watson's judgment. They measure the willingness of individuals at different income levels to pay for cleaner air and water. Watson judged that as income rises over time, individual's increased satisfaction (marginal utility) per unit of improvement in environmental quality would decline, even as total satisfaction with a cleaner environment increased. Watson, therefore, assumed declining income elasticities of environmental quality as follows: from 1975 to 1985 an income elasticity of 1.6 was used and between 1985 and 2000 an income elasticity of 0.67 was used.

Pollution control costs were calculated using the ABATE module of SEAS. The ABATE module is identical in concept to the energy investment module previously described. Instead of allocating energy investment costs to input-output sectors, it allocates investment in pollution-control equipment to these sectors.

The BENEFITS module calculated damages and control costs on an annualized basis and discounted control costs, damages, and benefits over the 1975 to 2000 period. The values provided are expected (that is, probability weighted) levels, based on an assumed probability distribution.

3 Input Assumptions

3.1 Solar-Energy Technologies: Generic Descriptions

Detailed engineering specifications for most solar systems are not relevant for this book because of the nature of the data input to the EIM of SEAS. In a given year, this data consists of total capital costs per delivered end-use energy for each solar-energy system, and a breakdown of these costs by input-output sector. As far as the calculation of the indirect environmental impacts of any particular solar technology is concerned, the model is blind to detailed engineering specifications. Two systems whose operational principles and actual designs differ markedly might have identical capital costs and sectoral breakdown of these costs and consequently have identical indirect environmental impacts.

3.1.1 Solar Heating and Cooling of Buildings

3.1.1.1 Passive System. The solar passive building design was provided by Hyde (1978). Two glazings of glass, each one-eighth inch thick, collect radiant solar energy, which is stored in an eighteen-inch thick concrete wall behind the glazings. No insulation is used. The generic system provides 66×10^6 Btu/yr or energy to the user (Hyde 1978).

3.1.1.2 Active System. The solar active heating and cooling system absorbs solar radiation by means of collectors, transfers the thermal energy to a transport medium or a working fluid, and then circulates the heated working fluid to a storage system. The stored thermal energy can be used at a later time for space heating and cooling and hot-water systems. The heating and cooling system that distributes the energy throughout the building is essentially identical to that used in conventional gas, oil, or electric systems.

An active solar heating system with double-glazed flat-plate collectors and liquid as a transport medium was considered along with an absorption air chiller (MITRE 1977a).

3.1.2 Solar Agricultural and Industrial Process Heat

A major use of thermal energy by the agricultural and industrial sectors is for relatively low temperature applications (less than 175°C). These applications include can washing, grain drying, plastic and concrete block curing, and metal finishing. Total-energy systems can also be applied to provide on-site electricity while utilizing waste heat for various purposes. Solar AIPH systems are suited to these applications, which are used largely in the day-time when sunlight is available. The three major generic systems used were: flat-plate AIPH hot-water system, parabolic-trough AIPH steam system, and parabolic-dish total-energy system.

3.1.2.1 Flat-Plate AIPH Hot-Water System. The major components employed in hot-water systems are similar to those in SHACOB systems. They include collectors, plumbing, pumps and controls, valves, a storage system, and heat exchangers. Single-glazed selective surface aluminum flat-plate collectors were considered. This system can obtain temperatures of about 80°C for AIPH applications (MITRE 1977a).

3.1.2.2 Parabolic-Trough AIPH Steam System. This system is based on the Acurex 3001 model and the Hexel Corporation Parabolic Trough design (Torkelson 1978). It is composed of parabolic-trough modules, electronically controlled on a single-axis tracking system. These collectors, oriented north-south, collect the sun's radiant heat, convert that energy to sensible heat in a working fluid, store the excess thermal energy, and distribute this sensible heat to an AIPH application. High temperatures (up to 315°C) can be attained using these parabolic trough collectors.

3.1.2.3 Parabolic-Dish Total-Energy System. The parabolic-dish system converts solar energy into both electricity and thermal energy. The system produces heat at a receiver located at the focal point of the parabolic dish. A system of this type is described in a report from the MITRE Corp. (1977a). The system used for costing purposes was the Omnium-G Solar-Powered Electrical Generating Plant with 6 meter diameter Model HTC-25 tracking concentrator. On a typical summer day in Anaheim, California, this system produces 20,000 Btu/m²/day of thermal energy and 570 Btu/m²/day of electrical energy (Zalinger, personal communication).

3.1.3 Solar-Thermal-Electric System

A solar-thermal-electric system collects, concentrates, and converts the sun's rays to heat energy and then to electricity by means of a steam Rankine conversion system. The central-receiver concept of solar-thermal-electric power systems was used in this book. The central-receiver system consists of a large field of two-axis tracking heliostats that concentrate solar energy on a tower-mounted receiver. Water circulates through the receiver and carries heat energy to an energy-conversion system and/or to a storage reservoir. A detailed description of the plant used as a model for this system is given in a report from Martin Marietta Corp. (1977).

3.1.4 Photovoltaics

The generic photovoltaic systems were developed by SERI. Both the dispersed and central power-station-photovoltaic-system designs utilize cadmium-sulfide (CdS) cells of 13 percent efficiency in a nontracking, nonconcentrating array with an 80 percent packing factor and a capacity factor of 0.21. CdS cells were chosen as an example of thin-film photovoltaic technology that might be utilized in the period up to 2000. The reader should not interpret this choice as a prediction that CdS will be the dominant photovoltaic technology during the period. Neither system includes storage, since storage is assumed to be more economical when applied to the utility grid rather than to individual plants. The arrays, therefore, are grid-coupled through a line-commutated inverter power-conditioning unit. Overall system efficiency is 10 percent in both cases.

3.1.4.1 Distributed System. The distributed system utilizes a roof-shingle module into which the photovoltaic array is built. In this system and the centralized system, a laminar surface is applied to protect the cell arrays from degradation. This system is sized to provide 70 percent of a small building's (that is, house, small office, store) electrical needs during worst-case insolation (January in the northern hemisphere). Electrical needs were assumed to include lighting and electrical appliances only. Heating and cooling needs are assumed to be met by solar-thermal applications. During summer months, the system will be a net exporter of electricity to the grid. The power company supplies electricity during dark hours and cloudy weather. The system power-conditioning unit uses a thyristor bridge with a line-commutated inverter and has a peak power rating of 5 kilowatts.

3.1.4.2 Central System. The central power system is a 1-MW_p module that can be coupled with other modules for larger plants. [(MW_p represents *peak* mega-watts, or the amount of capacity that would be obtained under maximum solar insolation (at noontime under cloudless skies)]. This system is similar in design to the distributed system except that the arrays are placed on support structures constructed specifically for the plants, rather than contained in roof shingles. Due to the higher voltage produced by the central power system, the power-conditioning unit relies on multipulse thyristor bridges, using series-coupled thyristors, to give the necessary voltages.

3.1.5 Wind Energy Conversion System

This system's cost and performance is based on a 1.5-MWe design of the type manufactured by General Electric and Kaman Aerospace Corporation (MITRE 1977a). A WECS plant of this size is most likely to be utilized by a municipal or public utility. A 1.5 MWe WECS provides 4,993 MWh/year of electricity for high wind regime (mean wind speed = 15 mph), and 4,336 MWh/year of electricity for moderate wind regime (mean wind speed = 13.7 mph).

3.1.6 Biomass Collection Systems

3.1.6.1 Manure Collection. No suitable description of this system was available. In this study, manure was considered to be a by-product of animal-husbandry operations. Costs of collection and residuals were assumed to be identical to those that would have existed if manure were collected for use as fertilizer rather than for use as fuel. Thus additional costs and direct and indirect residuals were not assumed to result directly from this system.

3.1.6.2 Agricultural and Forest Residue Collection. No detailed system description was available for collection of agricultural and forest residues. It was assumed that collection operations closely paralleled agricultural and forestry operations.

3.1.6.3 Silvicultural Farms. The silvicultural-farm design was taken from Gauthier (1978). A site area of forty acres in southeastern United States was assumed, providing an annual wood (oak or hickory) output of 17×10^6 Btu per oven dry ton of wood.

3.1.7. Biomass Conversion Systems

3.1.7.1 Anaerobic Digestion of Manure. The anaerobic-digestion model plant was taken from Ballou (1978) and was originally designed by Bio Gas of Colorado, Inc. A digestion capacity of 737,500 ft^3, provided by three primary digesters 100 feet in diameter by 31 feet deep, allows 147,500 pounds of volatile solids to be digested per day. At 5.9 pounds of volatile solids per head of cattle per day, 25,000 head of cattle are required to supply this energy. Gross energy production in the form of low Btu gas (71 percent CH_4; 29 percent CO_2—710 Btu/scf) from the system is 1.5×10^{11} Btu/yr.

3.1.7.2 Pyrolysis of Agricultural Residues. Pyrolysis is performed by the Purox System, designed by Union Carbide Corporation, and described in Fisher et al. (1976). The plant processes 700 tons/day of barley and cotton-field wastes and cotton-gin trash, and operates 330 days per year. Medium Btu gas is produced by the thermal-decomposition process, with a net energy value of 2.5×10^{12} annually.

3.1.7.3 Wood-Fired Steam-Electric Power Plant. A wood-fired steam-electric–generating plant is analogous to a coal-fired steam-electric plant except that dry wood chips are used instead of coal in the furnace. This plant consists of five equipment groups: front-end wood-handling equipment, the boiler plant, the steam turbine plant, the electric generation plant, and the miscellaneous plant group that supports the others (MITRE 1977b). A wood-burning furnace generates steam in a boiler with 1,000 psi at 1,000°F. The steam is used to generate electricity through a steam Rankine conversion system. This plant, with installed capacity of 46 MWe, requires 850 dry tons/day of wood chips.

3.1.7.4 Wood Stoves. The generic design for wood stoves was based on the Defiant® stove, designed and marketed by Vermont Castings, Inc. (1976). The Defiant ® design produces a horizontal flame flow and uses a secondary combustion chamber and a smoke passage that circles around the combustion chambers. This reduces overall heat loss to the flue, thereby increasing efficiency. Overall stove efficiency is 50 percent. Heat output depends on stove sizes and temperatures. The stove chosen for analysis has a maximum output of approximately 55,000 Btu/hr. A heating value of 17×10^6 Btu per over-dry ton of wood was assumed.

3.2 Solar-Energy Scenarios

In developing solar-energy scenarios the DPR assumed that a serious social and environmental constraint existed for coal production above 40 quads/yr (triple today's production) and above 15 quads/yr for nuclear energy. The DPR assumed a low probability for the commercial availability of a plutonium-based breeder reactor by the year 2000, and an insignificant probability for availability of a fusion system by the year 2000.

In the base case, implementation of the National Energy Act, aggressive solar research and development ($1 billion/yr) and continued popular and Congressional support and acceptance of solar energy were assumed. The DPR assumed that the combination of public support and energy prices at approximately $25/bbl would cause many of the solar technologies to

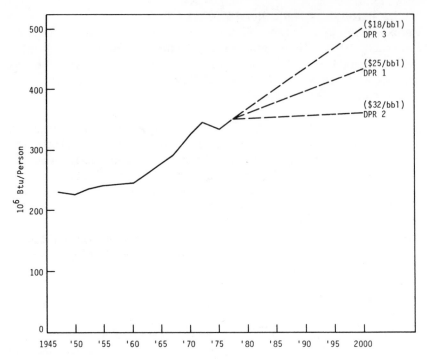

Sources: Economic Report of the President; Statistical Abstracts of the United States, 1978 .
Figure 3–1. Historic Rate of Per-Capita Energy Use

become attractive to society. Mature commercial production of solar
systems and the resolution of all the institutional, legal, and infrastructure
problems that currently face solar energy were also assumed.

The total domestic energy demand for the year 2000 was projected by
the DPR at 95, 114, and 132 quads for the high-, medium-, and low-price
scenarios, respectively. The estimated real growth rates in GNP associated
with these scenarios are 2.9 percent (95 quads), 3.1 percent (114 quads), and
3.3 percent (132 quads). These values were displayed in table 2–2 along with
the total estimated GNP (1977 dollars) and the ratio of total energy demand
to GNP. In this book only the $25/barrel (114 quads) case was analyzed.
The relationship between the assumed values for GNP and energy use and
historic trends and future projections is shown in figures 3–1 and 3–2. In all
scenarios, U.S. population of 262 million persons was assumed for the year
2000.

In developing the base-case scenarios, total energy demand—including
electricity along with its associated central-station generation losses—was

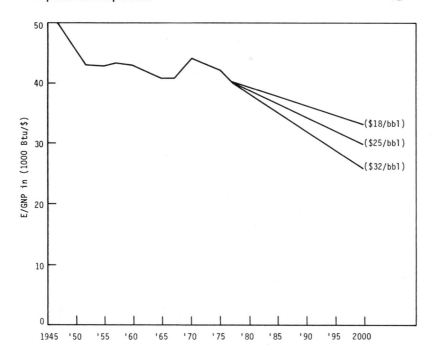

Sources: Economic Report of the President; Statistical Abstracts of the United States, 1978.

Figure 3–2. Historic Ratio of Total Energy Demand to GNP

estimated for the residential and commercial, industrial, and transportation sectors. Historical trends and future projections of per-capita use in each of these sectors are presented in figures 3–3, 3–4, and 3–5. In each case, sectoral energy demands were adjusted so that the total demand was equal to the values in table 3–1. The historical trend in electric power production is displayed in figure 3–6. A continuation of the trend toward electrification of the economy was assumed in the base case. The final breakdown in both demand and supply for the base-case scenarios is summarized in table 3–2 and displayed in figure 3–7.

All energy supply units in this analysis are fossil-fuel equivalents. A 10,400 Btu/kW heat rate (32.8 percent first-law conversion efficiency) was used for the first quad of equivalent energy displaced in a major energy sector such as the residential and commercial sector, the assumption being that initial uses of solar energy will substitute for electric resistance heating. Beyond 1 quad, the substitution ratio assumed a mixture of fossil combustion and electric-heat pump systems. Each quad of conventional energy was assumed to be replaced by 0.6 quads of solar energy.

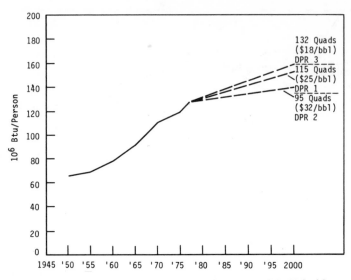

Sources: Economic Report of the President; Statistical Abstracts of the United States, 1978.
Figure 3–3. Residential and Commercial Per-Capita Energy Use

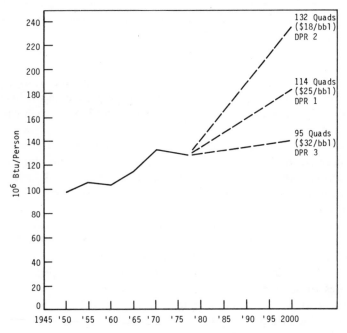

Sources: Economic Report of the President; Statistical Abstracts of the United States, 1978.
Figure 3–4. Industrial Per-Capita Energy Use

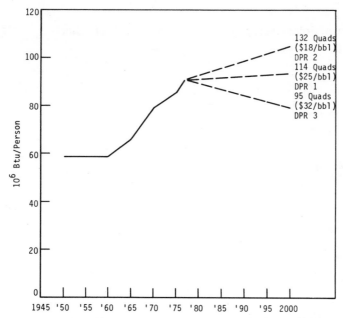

Sources: Economic Report of the President; Statistical Abstracts of the United States, 1978.

Figure 3–5. Transportation Per-Capita Energy Use

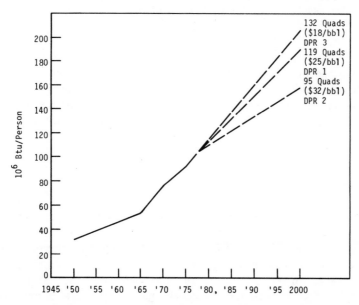

Sources: Economic Report of the President; Statistical Abstracts of the United States, 1978.

Figure 3–6. Electricity Per-Capita Energy Use

Table 3-1
Summary of Solar Energy in Base Cases by Year 2000, Quads

	Base Case		
	(DPR 3) *$18/bbl*	*(DPR 1, SERI 1)* *$25/bbl*	*(DPR 2)* *$32/bbl*
Hot water, heating and			
cooling, total	.6	1.1	1.6
Active	0.5	0.9	1.3
Passive	0.1	0.2	0.3
Industrial and agricultural			
process heat	0.2	1.0	1.4
Electricity, total	0.3	0.8	1.3
Wind	0.3	0.6	0.9
Solar thermal	—	0.1	0.2
Photovoltaic	—	0.1	0.2
OTEC	—	—	—
Biomass, total	2.4	3.1	4.4
Residential	—	0.1	0.1
Process heat	2.0	2.3	2.8
Electricity	0.1	0.2	0.6
Animal wastes	0.1	0.2	0.3
Synthetic fuels	0.2	0.3	0.6
Hydro electric[a]	3.3	3.3	3.3
Total	6.8	9.3	12.0
(%)	(5)	(8)	(12.6)

[a]The final DPR report may increase hydro by 0.5 to 0.7 quad.

The proper deployment of many solar-energy technologies requires energy storage and conventional backup. Storage assumptions vary greatly by technology and are implicit in the descriptions in section 3.1. The conventional backup required for solar-energy systems was not calculated separately but is implicit in the total-energy-demand assumptions used in the DPR energy scenarios.

3.2.1 The Solar Contribution to the Base Case

The DPR defined solar energy to include the following applications: active and passive heating and cooling of buildings, solar hot-water heating, agricultural and industrial process heat, electricity from solar-thermal conversion and photovoltaic cells, energy from biomass, energy from wind, ocean thermal-energy conversion, electricity from solar satellites, and hydroelectricity. With this convention, solar technologies—in the form of hydro dams and the burning of wood and forest residues—now displace about 4.5 quads of primary fuels or about 6 percent of the present national supply.

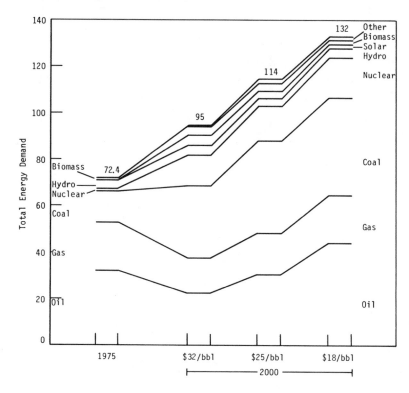

Figure 3-7. Energy Sources, Base Cases

Table 3-2 summarizes the contributions from solar energy in the base case in the year 2000. Solar energy, broadly defined, increases from about 4.5 quads (6 percent) at present to 9.3 quads (8 percent) in 2000, or by 107 percent of the present-day use.

3.2.2 High Solar-Energy-Use Scenarios

Two other energy scenarios used in this study were generated by the DPR based on the midprice base case but projecting greater use of solar energy. The maximum-practical case (DPR 4, SERI 2) would result from a determined effort at federal, state, and local levels to introduce solar technology into society. This scenario is shown in table 3-3. The total solar contribution is 16.8 quads or 15 percent of the energy supply by 2000. The fifth scenario generated by the DPR panel is the maximum-feasible or *technical limits* to solar-energy-use case (SERI 3). This case would result from all-out mobilization to increase solar-energy use, and the nation's physical produc-

Table 3–2
Base Case (DPR 1, SERI 1): Mid Price: U.S. Energy Demand and Supply for Year 2000, Quads
(*$25/bbl, 1978 dollars*)

| | End Uses | | | | | Intermediate Uses | | |
| | Residential and Commercial | | Industrial | | | | | |
Uses	Buildings[a]	Other	Process Heat	Other	Transportation	Electric Utility	Synthetic Fuels	Gross Energy Use
Primary								
Other	—	—	—	—	—	0.5	—	0.5
Oil	2.1	—	1.9	4.2	21.6	2.6	—	32.4
Gas	7.0	—	5.6	3.6	0.8	1.0	—	18.0
Coal	—	—	4.1	4.1	—	26.6	4	38.8
Nuclear	—	—	—	—	—	15.0	—	15.0
Hydro	—	—	—	—	—	3.3	—	3.3
Solar	1.1	—	1.0	—	—	0.8	0.2	3.1
Biomass	0.1	—	2.3	—	—	0.2	0.3[b]	2.9
Intermediate								
Electricity	18.4	11.8	1.8	17.6	1.0	50.6	—	—
Synthetic fuels	0.5	—	0.6	1.2	1.6	0.6	4.5	—
Total	29.2	11.8	17.3	30.7	25.0	—	—	—

[a] Space heating and cooling and water heating.
[b] Animal wastes may be considered biomass.

Table 3-3
Maximum-Practical Case (DPR 4, SERI 2): U.S. Energy Demand and Supply for Year 2000, Quads
($25/bbl, 1978 dollars)

| | End Uses | | | | | Intermediate Uses | | |
| | Residential and Commercial | | Industrial | | | | | |
Uses	Buildings[a]	Other	Process Heat	Other	Transportation	Electric Utility	Synthetic Fuels	Gross Energy Use
Primary								
Other	—	—	—	—	—	0.8	—	0.80
Oil	0.7	—	1.3	4.0	21.6	2.4	—	30.0
Gas	6.3	—	5.6	3.6	0.8	1.0	—	17.3
Coal	—	—	3.5	4.1	—	25.1	3.4	36.1
Nuclear	—	—	—	—	—	13.0	—	13.0
Hydro	—	—	—	—	—	3.9	—	3.9
Solar	3.0	—	2.0	—	—	3.4	0.3	8.7
Biomass	0.3	—	2.5	—	—	0.5	0.9	4.2
Intermediate								
Electricity	18.4	11.8	1.8	17.6	1.0	50.6	—	—
Synthetic fuels	0.5	—	0.6	1.4	1.6	0.5	4.6	—
Total	29.2	11.8	17.3	30.7	25.0	—	—	114

[a] Space heating and cooling and water heating.

Table 3–4
Maximum-Feasible Case (DPR 5, SERI 3): U.S. Energy Demand and Supply for Year 2000, Quads
($25/bbl, 1978 dollars)

| | End Uses | | | | | Intermediate Uses | | Gross Energy Use |
| | Residential and Commercial | | Industrial | | | | | |
Uses	Buildings[a]	Other	Process Heat	Other	Transportation	Electric Utility	Synthetic Fuels	Gross Energy Use
Primary								
Other	—	—	—	—	—	1.0	—	1.0
Oil	0.5	—	0.7	3.7	21.5	0.4	—	26.8
Gas	4.4	—	5.2	3.6	0.7	2.0	—	15.9
Coal	—	—	2.8	4.1	—	24.9	2	33.8
Nuclear	—	—	—	—	—	9.0	—	9.0
Hydro	—	—	—	—	—	4.0	—	4.0
Solar	5.0	—	3.5	—	—	8.0	0.5	17.0
Biomass	0.5	—	2.6	—	—	0.8	2.6	6.5
Intermediate								
Electricity	18.4	11.8	1.8	17.6	1.0	50.6	—	—
Synthetic Fuels	0.4	—	0.7	1.7	1.8	0.5	5.1	—
Total	29.2	11.8	17.3	30.7	25.0	—	—	114

[a] Space heating and cooling and water heating.

tion and installation capacity would be the major limits on solar-energy use. Market economics or other financial and institutional barriers would not constrain solar-energy use in any manner in the maximum-feasible case. The results of this exercise are shown in table 3–4. Total solar-energy use is 27.5 quads (24 percent) in 2000. The DPR included OTEC and solar satellites while the present study does not. This has no effect in the base case, since the DPR case from which this is taken did not specify any production of energy from OTEC or satellites by the year 2000. Even in the maximum-practical and maximum-feasible cases the production of energy from OTEC is minor by the year 2000: only .1 and 1.0 quads, respectively. In these cases the DPR projections of energy from these technologies were allocated to the remaining solar-electric options in proportion to their contributions in the year 2000. Satellites make no contribution by the year 2000 in any of the DPR scenarios. The solar-energy contributions in the year 2000 in the maximum-practical and maximum-feasible scenarios are summarized in table 3–5.

Once the solar-energy contributions for the high solar-energy scenarios were projected by the DPR, conventional fuel use was estimated. The general approach assumed that high solar-energy use will be caused primarily by a broadly based political decision to make a transition to solar energy.

Table 3–5
Summary of Solar Energy in Maximum-Practical and Maximum-Feasible Cases, Year 2000, Quads

Area	Maximum Practical	Maximum Feasible
Hot water, heating and cooling		
Active	2.0	4.0
Passive	1.0	1.0
Agricultural and industrial process heat	2.0	3.5
Electricity, Total	3.4	8.0
Wind	1.5	3.0
Solar thermal	0.8	1.5
Photovoltaic	1.0	2.5
OTEC	0.1	1.0
Biomass, total	4.5	7.0
Residential	0.3	0.5
Process heat	2.5	2.6
Electricity	0.5	0.8
Animal wastes	0.3	0.5
Synthetic Fuels	0.9	2.6
Hydro electric	3.9	4.0
Total, quads	16.8	27.5
(% of total energy supply)	(14.7)	(24)

Table 3-6
Assumed Regional Allocation of Solar Energy for All Scenarios, Year 2000
(percentage)

Solar Technology	New England [a] I	N.Y.-N.J. II	Middle Atlantic III	South-east IV	Great Lakes V	South Central VI	Central VII	Mountain VIII	West IX	North-west X
Utilities—wind	7	11	12	31	21	8	2	2	5	3
Utilities—photovoltaic	0	0	6	29	0	30	0	7	17	10
Utilities—solar thermal	0	0	4	21	6	19	8	12	21	8
Utilities—biomass	2	2	6	35	9	19	5	1	10	10
AIPH—total energy system	3	8	8	15	20	28	4	2	10	3
AIPH—flat-plate system	3	8	8	15	20	4	2	10	3	
AIPH—parabolic trough	3	8	8	15	20	28	4	2	10	3
Biomass—process heat	3	1	6	32	7	14	2	3	10	22
Animal residues	1	1	3	13	7	7	20	10	24	4
Agricultural and forest residues	3	1	5	31	4	22	2	2	9	20
Biomass farms	2	1	3	18	23	11	27	9	1	5
Anaerobic digestion	1	1	3	13	17	7	20	10	24	4
Pyrolysis	2	2	6	35	10	19	5	1	10	10
Photovoltaics—residential	0	0	6	29	0	31	0	7	17	10
SHACOB—active	4	14	8	9	27	14	6	4	11	3
SHACOB—passive	4	15	8	9	14	27	6	4	10	3
Wood stoves	16	7	10	7	35	0	1	14	0	11

[a] These regions are numbered according to standard federal procedure. The regions do not have official names; the names used here are for convenience only.

The social motivation for increased solar-energy market penetration was assumed to be both a reaction to the adverse characteristics of conventional energy systems and a positive reaction to some of the characteristics of solar energy.

Crude judgment was made by the DPR Impacts Panel that oil, coal, and nuclear energy are more similar in the magnitude of social disruption and difficulty than they are different and that gas has less serious social problems. In the year 2000, solar energy, therefore, was assumed by the DPR panel to substitute for coal, oil, nuclear energy, and gas in the following proportions: .3, .3, .3, and .1, respectively.

To determine the solar energy contributions in each scenario between 1975 and the year 2000, a growth rate of 35 percent per year for most of the solar technologies was assumed by SERI based on work done by Edelson and Lee (1976). Year 2000 solar contributions were then extrapolated backwards in time at -35 percent per year. The regional allocation of solar-energy use is shown in table 3–6.

3.3 GNP, Labor Productivity, and Employment

For an energy-economic scenario to make sense, its energy demand forecasts must be consistent with its GNP forecasts. Further, the GNP must be consistent with employment and productivity forecasts. All scenarios assumed a total energy supply of 114 quads in the year 2000. This was based on an assumed oil price of $25/barrel (1978 dollars) in the year 2000. These scenarios were devised by the President's Domestic Policy Review of Solar Energy group. This group, in consultation with Mike Maier of the Department of Commerce, determined that the GNP for the year 2000, consistent with this forecast, was $2687.42 billion (1972 dollars). Interim-year GNP targets, also in 1972 dollars, were: 1985, $1867.0 billion, and 1990, $2149.35 billion. These were determined by scaling the change in 1975–2000 GNP in the DPR scenarios to the GNP changes in the National Energy Plan (NEP) scenario previously run by SEAS. The GNP changes for the NEP scenario were obtained from the Office of Planning Analysis and Evaluation, Department of Energy. The increases in GNP in the DPR scenarios correspond to a 3.0 percent annual real exponential growth rate. This corresponds to a postwar real growth rate of 3.6 percent (Council of Economic Advisors 1975).

The GNP for other years was determined using linear interpolation. All scenarios assumed the following rates of unemployment: 1985, 5.0 percent; 1990, 4.8 percent; 2000, 4.8 percent.

4 Conclusions

4.1 Scenario Analysis

In this chapter, the scenarios described in section 3.2 are labeled as follows: the base case is labeled scenario 1, the maximum-practical case, scenario 2; and the maximum-feasible case, scenario 3. The letters H or L, which follow the number of the scenario, indicate high (H) or low (L) levels of environmental controls, as discussed in section 2.1.4.2. All references to environmental impact, environmental quality, and so on, should be interpreted as references to the production of air, water, and other residuals. The economic damages that result from these residuals are discussed in section 4.1.4.

4.1.1 Summary

On a national basis, rapid deployment of solar energy makes a major net-positive contribution to environmental quality in the year 2000. The effect, however, is less dramatic than casual reasoning might lead one to suspect. The reasons are the use of wood stoves, which emit large amounts of particulates to the atmosphere; silvicultural farming practices resulting in agricultural runoff problems; and the increased particulate emissions associated with the large increase in the size of the stone-and-clay-products industry that is required to produce some solar-energy systems. In spite of the beneficial environmental impact of solar-energy systems, if one wants to control environmental pollution, then environmental controls are a far more effective means than the deployment of solar energy.

The reduction in CO_2 emissions achievable by the year 2000 under even the maximum-feasible solar-energy scenario is negligible compared with the global atmospheric inventory of CO_2.

4.1.2 National Environmental Comparisons

Before analyzing the specific scenarios, several methodological points should be summarized. First, target GNP was held constant across all scenarios. Thus the effects observed here are independent of the macroeconomic (multiplier) effects that solar-energy deployment might have. (On the other hand, changes in the composition of GNP are accounted for.) GNP was held constant across scenarios to isolate any environmental effect specifically due to solar deployment.

Another consideration is that for the deployment of all solar technologies, except process heat from biomass (an already established technology), all the scenarios assumed a 35 percent per year deployment growth rate leading to the year 2000 solar-energy deployment figures. Since the solar-energy technologies are generally materials intensive (and thus residuals intensive) to build but are often benign to operate, an exponential growth rate in their construction tends to exaggerate the residuals that would be produced by large-scale use of solar energy in the long term. That is, the environmental improvement one might expect from the use of large amounts of solar energy is partially delayed by rapid growth in its use.

In this section, all water-residuals projections and comparisons are point source only. Section 4.1.4, however, is based on total point and non-point water residuals. Air residuals in both sections include transportation (non-point) sources.

In this book, comparisons between the base case and high solar scenarios are paramount. Important residuals for the maximum-practical case (2H) are compared in table 4-1 with those in the base case (1H), in the year 2000, to determine the relative impact of solar energy on the environment. For most residuals, differences of 2 percent or more are explored for their causes. Changes of 1 percent or more in the emission of particulates SO_x and NO_x are explored for their causes because of the relatively high damages associated with these pollutants. Table 4-1 shows the nationally aggregated impact of solar energy on the production of residuals. For water pollutants these, and subsequent calculations were made using data on point-source residuals only, since non-point-source water residuals were essentially identical for all scenarios. In none of these comparisons is it possible to ascertain which differences are statistically significant and which are not. Unfortunately, there is no information available on the distribution of errors associated with the results obtained using the SEAS model.

In the maximum-practical case (2H) particulates are 0.7 percent higher than in the base case (1H), while both SO_x and NO_x are 6 percent lower. Particulates increase primarily because of the increased use of wood stoves and cement and stone and clay products used particularly in the construction of passive solar systems but also in other solar-energy systems. Relative to the net increase of 345,000 tons of particulates, wood stoves contribute 70,000 tons and stone and clay products account for a net increase of 198,000 tons. These contributions do not add to the net increase in particulates due to offsetting decreases from other sectors of the economy. These are the only major contributing sectors. Table 4-2 shows the sources of absolute changes in residuals from the base case to the maximum-practical

Table 4–1
Comparison of the Base, Maximum-Practical, and Maximum-Feasible Cases for the Year 2000: National
(ratios of scenarios results)

	2H/1H	3H/1H	3H/2H
Particulates	1.007	1.016	1.009
SO_x	0.937	0.882	0.941
NO_x	0.939	0.908	0.967
HC	0.983	0.966	0.983
CO	1.000	1.003	1.003
CO_2	0.926	0.851	0.919
BOD	0.997	0.993	0.997
COD	1.048	1.163	1.110
Suspended solids	1.128	1.446	1.282
Dissolved solids	0.948	0.887	0.936
Solid wastes	0.968	0.992	1.024
Industrial sludges	0.902	0.891	0.988
Water use	0.973	0.929	0.955
Permanent land use	0.996	0.977	0.981
Temporary land use	1.530	2.914	1.904
Injuries	0.999	0.996	0.996
Workdays lost	0.997	0.993	0.996
Operational labor	1.077	1.245	1.156

1H = Base-case emissions (high environmental control).
2H = Maximum-practical case emissions (high environmental control).
3H = Maximum-feasible case emissions (high environmental control).

case. As might be expected, the decreases in SO_x and NO_x result entirely from the decreased combustion of fossil fuels. This also is shown in table 4–2. HC and CO show less than a 2 percent difference between scenarios, while CO_2 is 8 percent lower in the maximum-practical case (2H). This is due to decreased use of fossil fuels. For the water residuals, suspended solids are almost 13 percent greater in scenario 2H (completely due to increases in silvicultural farming) while industrial sludges are 10 percent lower due to decreased use of fossil energy. Less important changes occur with COD, dissolved solids, and water use, which are 3 percent greater, 5 percent less, and 3 percent less in the maximum-practical case (2H), respectively. The increase in COD in the maximum-practical case is almost entirely due to the increased use of AIPH systems, which emit a great deal of propylene or ethylene glycol (whose degradation is oxygen demanding). Dissolved solids decrease due to decreased use of fossil and nuclear fuels.

Table 4–2
Sectors Causing Changes in National Residuals Outputs between the Maximum-Practical Scenario and the Base Case for the Year 2000
(tons unless noted)

Aggregated Sectors	Partic-ulates	SO_x	NO_x	HC	CO	BOD	COD
Municipal sewage and waste	−1810	−1077	−1595	−1385	−8855	0	0
Agriculture	0	0	0	0	0	0	0
Silviculture	0	0	0	0	0	0	0
Feedlots and fisheries	0	0	0	0	0	0	0
Food products processing	−168	0	0	0	0	−571	−490
Textiles	0	0	0	0	0	−192	−1399
Leather goods	0	0	0	0	0	−17	−48
Pulp and paper	−629	−489	0	0	−6448	−1374	0
Lumber products	0	0	0	−613	0	0	0
Inorganic chemicals	−552	−2959	−98	0	0	0	2
Organic chemicals	−43	−591	−54	−3214	−3335	−168	−4306
Drugs	0	0	0	0	0	−85	−206
Fertilizers	−54	0	−217	0	0	0	0
Glass	0	0	0	0	0	−3	−6
Plastics	−16	0	0	−164	0	−92	−324
Fiberglass	−1	−26	−31	0	0	0	0
Synthetic rubber	0	0	0	0	0	−25	−275
Soaps, detergents	−596	0	0	0	0	−48	−200
Paints	0	0	0	−2715	0	0	0
Asphalt	−4304	0	0	0	−5069	0	0
Stone and clay products	142436	14646	36784	−4	−615	1	3
Cement	8974	15041	36784	0	0	0	0
Other structural materials	189262	−396	0	−4	−615	1	3
Iron and steel	1316	−30	−1	120	77973	0	0
Nonferrous metals	26092	6087	0	0	0	0	65
Copper	80	6194	0	0	0	0	0
Aluminum	26002	0	0	0	0	0	63
Other	10	−108	0	0	0	0	2
Metal fabrication	0	0	0	0	0	0	0
Electroplating	0	0	0	0	0	0	0
Printing	0	0	0	−262	0	0	0
Service industries	0	0	0	−587	0	0	0
Medical services	0	0	0	0	0	0	−2
Electric utilities	−28523	−764723	−758020	5127	−38595	−925	−57978
Utilities coal—old	0	0	0	0	0	0	0
Utilities coal—new	−39460	−523983	−455930	−9015	−30674	−712	−69185
BACT	−33261	−463574	−383459	−8270	−27545	−614	−59615

Table 4-2 continued

Aggregated Sectors	Partic-ulates	SO_x	NO_x	HC	CO	BOD	COD
NSPS	− 5630	− 58135	− 38875	− 667	− 2221	− 49	− 4806
Fluidized bed and combustion cycle	− 570	− 2275	− 33596	− 78	− 908	− 49	− 4764
Utilities—oil	9260	− 241588	− 309935	2838	− 11274	− 151	− 13367
Utilities—gas	− 23	− 15	− 1673	− 2	− 41	− 3	− 259
Utilites—geothermal	0	0	0	0	0	0	0
Utilities—nuclear	0	0	0	0	0	− 59	− 1741
Utilities—solar	1697	849	9517	11305	3395	0	26574
Wind	0	0	0	0	0	0	0
Photovoltaic	0	0	0	0	0	0	0
Solar thermal	0	0	0	0	0	0	26574
Biomass	1697	849	9517	11305	3395	0	0
Industrial energy use	− 62410	− 496893	− 384648	− 10038	− 19572	0	116701
Coal	− 26055	− 325655	− 237545	− 3575	− 11493	0	0
Coal—old	− 8509	− 64978	− 20678	− 322	− 1033	0	0
Coal—new	− 17545	− 260677	− 216867	− 3254	− 10459	0	0
Oil and gas	− 36355	− 171237	− 147103	− 6463	− 8079	0	0
Solar	0	0	0	0	0	0	116701
AIPH—total-energy system	0	0	0	0	0	0	30246
AIPH—flat-plate	0	0	0	0	0	0	5893
AIPH—parabolic trough	0	0	0	0	0	0	80563
Residential—fossil	− 30791	− 81137	− 52726	− 10394	− 18865	0	0
Commercial—fossil	− 35967	− 218989	− 115548	− 5901	− 8496	0	0
Residential and Commercial—solar	343230	0	0	0	0	0	71789
Biomass	343230	0	0	0	0	0	0
Active	0	0	0	0	0	0	71789
Passive	0	0	0	0	0	0	0
Electric—photovoltaic	0	0	0	0	0	0	0
Transportation	1784	215	219	9357	116112	0	0
Coal mining	− 1597	− 205	− 2807	− 281	− 1737	0	0
Coal processing	− 1077	− 6	− 713	− 281	− 200	0	0
Coal transportation	− 37441	0	0	0	0	0	0
Low Btu gas	0	0	0	0	0	0	0
High Btu gas	− 54	− 761	− 215	− 5	− 17	− 29	0
H—coal	− 2113	− 5911	− 13340	− 201	0	− 49	− 212
Oil shale	− 436	− 817	− 714	− 155	0	0	0
Gas processing, distribution, extraction	− 170	− 135	− 8125	− 622	− 2338	0	0
Oil distribution and extraction	− 7618	− 9389	− 81707	− 87391	− 26580	0	0
Petroleum refining	− 10953	− 98302	− 45208	− 41925	− 3498	− 552	− 1559
Nuclear processing and distribution	0	− 294	− 237	− 16	− 1	0	0

Table 4–2 continued

Aggregated Sectors	Partic- ulates	SO_x	NO_x	HC	CO	BOD	COD
Biomass collection	458	1054	7904	1722	10863	0	0
Agricultural and forest residue	221	589	4419	958	6077	0	0
Biomass—farms— wood	237	465	3485	765	4786	0	0
Synthetic gas—biomass	394	6736	6382	98	984	0	0
Animal residue	0	4965	4413	0	0	0	0
Agricultura and forest residue and waste	394	1772	1968	98	984	0	0
Subsectors in groupings	344182	−1654000	−1414688	−149728	61712	−4044	121761
Subsectors not in groupings	335	0	0	−10291	0	−1	−6
Grand total	288717	−1654000	−1414688	−160019	61712	−4046	121755

Note: For water residuals, point source only, is given. All numbers in the table represent differences between scenarios, not absolute values. Negative signs indicate a decrease compared with the base case.

The decreased use of water in the maximum-practical case is due only to the decreased use of cooling water for nuclear power plants, which is partially offset by increased water use for solar-thermal power plants and on silvicultural farms. BOD shows virtually no variation between scenarios. Solid wastes are 3 percent lower in the maximum-practical case than in the base case. Permanent land use is unchanged, while temporary land use is 53 percent greater in the maximum-practical case; injuries and workdays lost remain the same. Solid waste decreases almost entirely due to the decrease in use of coal and oil. The decrease in solid wastes associated with reductions in the use of these fossil fuels is in large part offset by increases in solid waste resulting from the increased use of stone and clay products for solar-energy construction and in silvicultural farming. The increase in land use is entirely due to increased silvicultural farming.

Originally, I conjectured that the maximum-practical case might show positive environmental benefits in the year 2000 but that the maximum-feasible case would result in a negative environmental impact due to the large amount of solar-related investment required by an all-out push to develop solar energy. This is not the case. The maximum-feasible case yields changes from the base case that are identical in sign but larger in magnitude than the maximum-practical case. The only residual for which this is not true is solid waste. While maximum-practical case solid wastes are 3 percent lower than in the base case, maximum-feasible case solid wastes are less than 1 percent smaller than those in the base case. This is because the large decrease in solid wastes in scenario 2, caused by reduced use of coal and oil in both scenarios, is largely offset in scenario 3 by the increased solid waste associated with production of synthetic gas by pyrolysis of agricultural and forestry wastes.

Maximum-feasible case (3H) particulates are 1.6 percent greater than in the base case with high environmental controls, 1H, while SO_x, NO_x, HC, and CO_2 are 12, 9, 3, and 15 percent lower, respectively. CO variations are again unimportant. Suspended solids and COD are 45 and 16 percent higher, respectively in scenario 3H, while dissolved solids and industrial sludges are both 11 percent lower, and water use is 7 percent lower. BOD, sewage sludge, and solid wastes do not vary significantly between scenarios. Permanent land use does not change, while temporary land use is almost 200 percent greater in the maximum-feasible case (due entirely to the increased use of silvicultural farming for fuel production). Injuries and workdays lost do not change significantly. Changes in residuals outputs from scenario 1H to scenario 3H have the same causes as the differences in residuals outputs between scenarios 1H and 2H.

Table 4-1, column 3, compares year 2000 residuals in the maximum-feasible case (3H) and the maximum-practical case (2H). The largest difference between the two scenarios occurs with temporary land use, which is 90 percent higher in 3H. Noteworthy differences also occur with particulates, COD, suspended solids, solid wastes, and operational labor, which are higher in scenario 3H by 4, 11, 28, 3, 13, and 16 percent, respectively. SO_x, NO_x, CO_2, dissolved solids, and water use are all lower in scenario 2H by 6, 3, 8, 6, and 4 percent respectively. The other residuals considered showed no major differences. Summarizing these results, with some exceptions, solar energy makes a positive contribution to environmental quality in the year 2000, though not as great a contribution as one might have expected before doing the analysis.

A question of interest is whether the imposition of strict environmental controls has a more or less important effect on the production of residuals than the introduction of large amounts of solar energy. This question was addressed because occasionally solar-energy deployment is put forth primarily as a mechanism for improving the environment. This question is also important because it provides a standard by which the magnitudes of solar-related residuals changes can be evaluated. This question can be analyzed by examining table 4-3, which compares the percentage changes in residuals occurring as the switch is made from the low-control base case to the low-control high solar cases with percentage changes in residuals occurring as the switch is made from the low-control–base case (1L) to the high-control–base case (1H). With a few exceptions EPA controls, used without solar energy, are projected to have a more significant and beneficial impact on the environment than deployment of solar energy without environmental controls, even in the maximum-feasible case. Particulates, HC, CO, COD, and suspended solids show decreases of 9 to 70 percent in scenario 1H, relative to scenario 1L. The same pollutants show increases of 1 to 32 percent, or no changes, in scenarios 2L (maximum-practical case–low control), and 3L (maximum-feasible case–low control), relative to 1L, (base case–low control). SO_x, NO_x, and BOD decrease by 13, 27, and 25 percent respectively, in scenario 1H compared to 1L. Corresponding decreases in scenario 3L are 13, 7, and 1 percent for SO_x, NO_x, and BOD, respectively. An exception to the generalization that solar energy has less beneficial environmental effects than pollution control is industrial sludges. These increase by 46 percent in the high-control–base case compared to the low-control–base case, while decreasing by 6 percent in the low-control–maximum-practical case (2L), and 8 percent in the low-control–maximum-feasible case (3L), compared to the low-control–base case. Another exception is dissolved solids, which decrease by similar amounts in scenarios 1H and 2L, compared to scenario 1L.

Another question of interest is whether the degree of environmental control has any influence on the relative effect of increasing use of solar energy. In other words, does the increased use of solar energy have a different impact under a set of stringent environmental controls than it might under less stringent controls? This question can be analyzed by looking at table 4-4, in which the percentage changes as one moves from scenario 1L to 2L are divided by the percentage changes as one moves from scenario 1H to 2H. Quotients in the neighborhood of one indicate little relative impact of environmental controls on the residuals impact of increased solar-energy use. For most residuals, the effect of environmental controls on the relative effect of increasing solar energy is small. In no case does this impact exceed 10 percent of the differential effect. In other words, while solar energy has a larger absolute environmental benefit when few environmental controls are in place, the relative effect of increased solar deployment is not significantly changed by the degree of environmental control.

Brief mention of the projected effects of solar energy on the emission of carbon dioxide (CO_2) should be made. These effects are potentially important because of the widespread suspicion that fossil-fuel-related CO_2 emissions may be inducing climatic change via the greenhouse effect. Base-case CO_2 emissions from the energy sector in the year 1975 are 4.1×10^9 tons. Base-case energy-sector CO_2 emissions in the year 2000 are projected to be 5×10^9 tons. Under the maximum-practical and maximum-feasible scenarios, energy-sector CO_2 emissions in the year 2000 are projected to be 4.7 and 4.3×10^9 tons, respectively, resulting in annual decreases of approximately

Table 4–3
Comparison of High Environmental Control Technology (Scenario 1H)
and Maximum-Practical and Maximum-Feasible Scenarios (2L and 3L)
with Low Control (1L) for the Year 2000: National
(ratios of scenario results)

	(1) *1H/1L*	*(2)* *2L/1L*	*(3)* *3L/1L*	*(4)* *1H/1L* *2L/1L*	*(5)* *1H/1L* *3L/1L*
Particulates	0.90	1.03	1.06	0.87	0.85
SO_x	0.87	0.93	0.87	0.94	1.00
NO_x	0.73	0.96	0.93	0.76	0.78
HC	0.48	0.99	0.99	0.48	0.48
CO_2	0.30	1.00	1.00	0.30	0.30
BOD	0.75	1.00	0.99	0.75	0.76
COD	0.74	1.03	1.12	0.72	0.66
Suspended solids	0.71	1.09	1.32	0.65	0.54
Dissolved solids	0.95	0.95	0.89	1.00	1.07
Solid wastes	0.99	0.98	1.01	1.01	0.98
Industrial sludges	1.46	0.94	0.92	1.55	1.59
Water use	1.01	0.97	0.93	1.04	1.09
Permanent land use	1.00	1.00	0.98	1.00	1.02
Temporary land use	1.00	1.53	2.92	0.65	0.34
Injuries	1.00	1.00	1.00	1.00	1.00
Workdays lost	1.00	1.00	0.99	1.00	1.01

Note: Column 1 indicates the effect of switching from low environmental controls to high environmental controls.
Columns 2 and 3 indicate the effects of switching from the low-control base case to the low-control maximum-practical or feasible cases.
Columns 4 and 5 indicate the relative effects of switching from low to high environmental controls versus switching from low to high deployment of solar energy.
1H = Base-case emissions (high environmental control).
1L = Base-case emissions (low environmental control).
2L = Maximum-practical case emissions (low environmental control).
3L = Maximum-feasible case emissions (low environmental control).

6 and 14 percent compared to the base case. Even a 14-percent decrease in U.S. energy-related CO_2 emissions is negligible on a world scale, however. CO_2 already in the atmosphere amounts to approximately 3×10^{12} tons; thus the change of $.7 \times 10^9$ tons of CO_2 emitted annually under the maximum-feasible scenario represents just over .02 percent of the current atmospheric inventory. Even if the CO_2 savings from the maximum-feasible scenario were maintained over a one-hundred-year period, the total savings would be only 2.5 percent of the global inventory (assuming there is no effect on the equilibrium CO_2 inventory). This compares with a 13-percent

Table 4–4
Relative Effects of Pollution Control on Scenarios in the Year 2000: National
(ratios of scenario results)

	(1) 2L/2H / 1L/1H	(2) 3L/3H / 1L/1H	(3) 3L/3H / 2L/2H
Particulates	0.99	0.99	1.00
SO_x	1.00	0.99	.99
NO_x	1.02	1.03	1.01
HC	1.01	1.02	1.01
CO_2	1.01	1.00	1.00
BOD	1.01	1.01	1.00
COD	0.99	0.96	0.97
Suspended solids	0.97	0.91	0.94
Dissolved solids	1.00	1.00	1.00
Solid wastes	1.01	1.01	1.00
Industrial sludges	1.04	1.03	1.00
Water use	1.00	1.01	1.01
Permanent land use	1.00	1.00	1.00
Temporary land use	1.00	1.00	1.00
Injuries	1.00	1.01	1.01
Workdays lost	1.00	1.00	1.00

Note: Column 1 compares the effect of switching from high to low environmental controls under the base case with the same effect under the maximum-practical case.
Column 2 compares the effect of switching from high to low environmental controls under the base case with the same effect under the maximum-feasible case.
Column 3 compares the effect of switching from high to low environmental controls under the maximum-practical case with the same effect under the maximum feasible case.
1H = Base-case emissions (high environmental control).
1L = Base-case emissions (low environmental control).
2L = Maximum-practical-case emissions (low environmental control).
3L = Maximum-feasible-case emissions (low environmental control).

increase in CO_2 levels over the past century (EPRI Journal 1978). Thus the use of solar energy in the United States alone, on a scale contemplated by the Domestic Policy Council in the year 2000, would have no significant effect on global CO_2 emissions. A far more significant reduction in energy-related CO_2 emissions could be achieved by limiting the growth of total energy demand to well below the Domestic Policy Council's projection of 114 quads in the year 2000. On the other hand, if the United States effected a significant transition to solar energy, so might other industrialized nations. Collectively over a long period of time the effect on CO_2 emissions could be significant.

4.1.3 Regional Environmental Comparisons

As part of this study, a comparison of the pollutant emission effects of the solar- and conventional-energy scenarios was made for each of the ten federal regions. It was anticipated that the extent of emissions reductions or increases would vary widely from region to region and the results of such a comparison would offer a guide to a regional and political analysis of the environmental benefits and costs of solar energy.

When the data were examined, no clear pattern of regional variations in emissions emerged. Thus the results are not reported here. Some information on the regional economics of solar energy's environmental benefits and costs is offered in section 4.1.4.

4.1.4 Economics of Environmental Comparisons

4.1.4.1 Summary. Estimates of the dollar value of the environmental benefits from deployment of solar energy are given here. These estimates were made using the experimental BENEFITS model described in section 2.3. The estimates cover the period from 1975 to 2000 and are presented in conventional present discounted-value terms. These estimates rely on the residuals estimates from both point and non–point sources. Thus all the caveats that apply to these residuals estimates apply equally to the economic estimates.

The methodology used to develop these estimates is an experimental one in need of further development. Moreover, both the economic and environmental data on which the methodology depends are under continuous development. The estimates presented here should be regarded as tentative in the extreme; more as indicators of magnitude and direction than as detailed projections.

The environmental benefits from deployment of solar energy at the maximum-practical level from 1975 to 2000 are estimated at $57.9 billion (1972 dollars), using an assumed value of a human life of $200,000 (1972 dollars) and a real discount rate of 2.5 percent. Estimates increase significantly if a higher value for human life is assumed and decrease significantly if a higher discount rate is employed in the calculations. The overwhelming majority of solar-energy's environmental benefits are estimated to stem from reductions in sulfur oxides associated with combustion of fossil fuels.

These estimates seriously understate the environmental benefits of solar energy for the following reasons:

Most important, the benefit stream from solar technologies is truncated in the year 2000. Significantly increased environmental benefits from solar-energy deployment are anticipated as the level of solar-energy use reaches its steady-state level in the twenty-first century;[1]

A conservative estimate of the value of a human life is used;

Analysis of benefits from pollutant reductions is limited to a small number of pollutants;

Stringent environmental controls on fossil fuel facilities are assumed, reducing the potential environmental benefits of solar energy technologies; and

Unsubstantiated health effects of some pollutants, such as carcinogenetic effects are largely ignored in the analysis.

4.1.4.2 Results. Table 4–5 summarizes the results of the environmental benefit-cost analysis of solar-energy scenarios. This analysis was prepared using the BENEFITS model described in section 2.3. The estimates are based on regionalized environmental-damage functions for the five criteria air pollutants and an index of water pollutants. Columns 4 and 5 of table 4–5 show the present discounted value of the net benefits from solar-energy scenarios compared to the base case over the 1975–2000 period. These results are portrayed as a function of the assumed value of life and assumed real rate of discount. All values in this section are in 1972 dollars.

Only the environmental benefits from high environmental-control solar-energy scenarios are analyzed. This use of the high-control scenarios was a deliberate attempt to understate the environmental benefits of solar energy.

Before examining the absolute magnitudes of the net environmental benefits from solar energy, several comparisons can be made. First, net benefits of the maximum-feasible case are always significantly greater than the net benefits of the maximum-practical case, compared to the base case, by a factor that ranges between 1.7 and 1.9. The increase in solar energy in the maximum-feasible case over the base case is 2.4 times as great as the increase in solar deployment in the maximum-practical case. Thus the environmental benefits from solar energy are, apparently, a decreasing nonlinear function of solar-energy deployment. This is to be expected since environmental damages are an increasing function of emissions in the BENEFITS model. Second, the net benefits of solar energy are a monotonically increasing function of the assumed value of human life in the environmental-health-effects calculations. This result is to be expected since a significant portion of national control total-environmental damages are related to health. The assumed values range from $30,000 (1972 dollars) to

Table 4-5
Comparison of Discounted Expected Environmental Damages, 1975 to 2000: Alternative Scenarios, Values of Life, Discount Rates
(billions of 1972 dollars)

Value of Life	(1) BC (1H)	(2) MPC (2H)	(3) MFC (3H)	(4) Net Benefits of MPC (1H-2H)	(5) Net Benefits of MFC (1H-3H)	(6) Ratio Net Benefits MFC/MPC
Real discount rate = 0.0%						
30,000	1473.8	1432.5	1403.0	41.3 (2.8)	70.8 (4.8)	1.7
100,000	2291.3	2226.9	2182.4	64.4 (2.8)	108.6 (4.8)	1.7
200,000	3459.4	3361.9	3295.7	97.5 (2.8)	163.7 (4.8)	1.7
300,000	4622.3	4492.1	4404.4	130.2 (2.8)	217.9 (4.8)	1.7
Real discount rate = 2.5%						
30,000	1076.0	1051.4	1033.1	24.6 (2.3)	42.9 (3.9)	1.7
100,000	1675.1	1636.8	1609.0	38.3 (2.3)	66.1 (3.9)	1.7
200,000	2531.1	2473.2	2431.8	57.9 (2.3)	99.3 (3.9)	1.7
300,000	3383.4	3306.0	3251.2	77.4 (2.3)	133.2 (3.9)	1.7
Real discount rate = 5.0%						
30,000	817.2	802.2	790.5	15.0 (1.8)	26.7 (3.3)	1.8
100,000	1273.7	1250.5	1232.7	23.2 (1.8)	41.0 (3.3)	1.8
200,000	1926.1	1890.9	1864.5	35.2 (1.8)	61.6 (3.3)	1.8
300,000	2575.6	2528.6	2493.5	47.0 (1.8)	82.1 (3.3)	1.8
Real discount rate = 7.5%						
30,000	643.3	634.0	626.4	9.3 (1.4)	16.9 (2.6)	1.8
100,000	1003.8	989.4	977.8	14.4 (1.4)	26.0 (2.6)	1.8
200,000	1519.0	1497.2	1479.9	21.8 (1.4)	39.1 (2.6)	1.8
300,000	2031.9	2002.8	1979.8	29.1 (1.4)	52.1 (2.6)	1.8
Real discount rate = 10.0%						
30,000	522.9	517.0	511.9	5.9 (1.1)	11.0 (2.1)	1.9
100,000	816.7	807.6	799.8	9.1 (1.1)	16.9 (2.1)	1.9
200,000	1236.6	1222.8	1211.2	13.8 (1.1)	25.4 (2.1)	1.8
300,000	1654.6	1636.2	1620.8	18.4 (1.1)	33.8 (2.1)	1.8

Note: ( ) indicates percent of base-case discounted damages.

$300,000 (1972 dollars) per human life. These figures span the conservative end of the range cited in the studies in Fisher (1979). Finally, the present value of net environmental benefits due to deployment of solar energy obviously declines as the real discount rate employed in the calculations rises. Real rates of discount of 0 to 10 percent are employed. At an assumed value of life of $300,000, the net environmental benefits of the maximum-feasible deployment of solar energy between 1975 and 2000 drop from $218 billion to $33.8 billion as the discount rate rises from 0 to 10 percent. A 0 percent real discount rate implies that environmental damages that accrue in the future are counted with equal weight as compared with present-day damages. This alternative is preferred by many environmentalists in benefit-cost calculations. It corresponds to a rate of interest equal to the rate of inflation or about 10 percent per year at present trends. A real rate of discount of 10 percent (or a rate of interest of about 20 percent given current trends) probably exceeds the pretax marginal productivity of capital, an often-used theoretical standard in social discounting.

A reasonable mid-range assumption for the value of human life is $200,000 (1972 dollars) or $292,000 (1977 dollars).[2] A reasonable real rate of discount is 2.5 percent. Using these mid-range assumptions, the net environmental benefits of the maximum-practical scenario are $57.9 billion (1972 dollars) and those of the maximum-feasible scenario, $99.3 billion.

Although the above estimates of the net environmental benefits of solar energy are large in absolute terms, it is necessary to have some standards of comparison to judge them on a national scale. One such standard is the size of environmental benefits compared with current-year GNP. GNP for 1978 (the last year for which figures were available at the time this work was completed) is $1,412.2 billion (1972 dollars) (Council of Economic Advisors, 1975, table B-7). Net environmental benefits of solar energy range from $218 billion to $5.9 billion. Thus, as a percentage of 1978 GNP, solar energy's discounted environmental benefits range from 15.4 to 0.4 percent, as the deployment of solar-energy discount rate and assumed value of life are varied.

The high end of this range represents an astoundingly large estimate of the net environmental benefits of solar energy between 1975 and 2000. This is the estimate obtained from maximum-feasible case, assuming a 0 percent real discount rate and a $300,000 value of life. Even the low estimate of solar's environmental benefits, 0.4 percent is approximately the size of the Department of Energy's annual budget in relation to GNP. The reasonable mid-range assumptions of $200,000 per life and a 2.5 percent real rate of discount work out to 4.0 percent and 7.0 percent of GNP for the maximum-practical and maximum-feasible scenarios, respectively. These are obviously quite large numbers, however, the environmental benefits from solar energy over a twenty-five year period are being compared with the current year's GNP.

Another measure of the size of net environmental benefits of solar energy is their size in relation to base-case total discounted environmental damages. These figures, shown in parentheses in columns 4 and 5, table 4–5, range from a high of 4.8 percent for the maximum-feasible case using a 0 percent real rate of discount to a low of 1.4 percent for the maximum-practical case using a discount rate of 10 percent. Thus in any given year the deployment of solar energy can mitigate between 1 percent and 5 percent of total environmental damages. Using the reasonable mid-range assumptions, maximum-practical and maximum-feasible solar deployments result in environmental benefits equal to 2.3 percent and 3.9 percent of environmental damages, respectively.

A third measure of the net environmental benefits of solar energy is their size per capita. Current U.S. population is approximately 218.5 million (Council of Economic Advisors, 1975, table B–26). Net environmental benefits range from a high of $218 billion for the maximum-feasible case using a $300,000 assumed value of human life and a 0 percent real rate of discount to a low of $5.9 billion using a $30,000 value of life and a 10 percent real rate of discount. This works out to a range of $1000 per capita to $27 per capita. Using the reasonable mid-range assumption of $200,000 per human life and a 2.5 percent real rate of discount, the net environmental benefits of the maximum-practical scenario are $266 per capita and the net environmental benefits of the maximum-feasible scenario are $454 per capita. These are substantial numbers to most individuals.

Fourth, one could examine the size of solar energy's environmental benefits in relation to the amount of solar energy used in any given year. For this analysis, it would be most appropriate to compare annual net benefits to annual solar-energy use, rather than comparing the present value of all future environmental benefits from solar energy to the amount of solar energy used in only one year. Proceeding on this basis, the environmental benefits generated by all forms of solar energy (including hydroelectric energy) can be compared with the amount of solar energy deployed in the year 2000. The net environmental benefits of the maximum-practical deployment of solar energy are calculated from table 4–12 at $13.4 billion ($109.8 billion to $96.4 billion) in the year 2000. The net increase in solar energy utilized in the maximum-practical case compared to the base case in the year 2000 can be calculated from tables 3–2 and 3–5 and is 7.5 quads. Solar energy's average environmental benefits are thus worth approximately $1.80/MMBtu (1972 dollars) or roughly $2.75/MMBtu (1978 dollars). At 5.8 MMBtu/bbl, this is equivalent to an environmental price of $15.95/bbl—over 60 percent of the $25/bbl (1978 dollars) which we expected to prevail in the year 2000. Similarly, solar energy's average environmental benefits in the maximum-feasible case are $1.54 (1978 dollars)/MMBtu of solar energy in the year 2000.

Finally, one could examine the net environmental benefits of solar energy per unit of energy currently utilized. The United States is currently utilizing roughly 78 quads per year of energy supply. At an average of 5.8 MMBtu/bbl, this is 13.4 billion bbl oil equivalent per year. Thus net environmental benefits of the solar-energy scenarios range from $16.20 per bbl oil equivalent per year to $.44 per bbl oil equivalent per year, with reasonable mid-range assumptions of a $200,000 value of human life and a 2.5 percent real rate of discount yielding net environmental benefits of $4.30 and $7.38 per bbl oil equivalent per year for the maximum-practical and maximum-feasible scenarios, respectively. Regardless of which standard of comparison is used, the environmental benefits of deploying solar energy between 1975 and 2000 appear quite substantial.

All of the above comparisons are shown in table 4–6. The results were aggregated from more detailed information on environmental damages associated with each scenario. In tables 4–7 through 4–10 the detailed information on environmental damages is broken down by major damaging sector (electric utilities, industrial point sources, residential and commercial fuels combustion, transportation, municipal waste water, urban runoff, agriculture, mining and other non–point sources). From these tables, one can calculate which sectors contribute most to the environmental benefits of solar-energy scenarios. Table 4–11 presents an exemplary calculation for the mid-range reasonable case of $200,000 assumed value of human life and 2.5 percent real rate of discount. It was constructed from table 4–9. Table 4–11 shows clearly that the overwhelming percentage (96 percent in the maximum-practical case) of the net environmental benefits from solar-energy deployment come from just three sectors: electric utilities, industrial point sources, and residential- and commercial-fuels combustion. Within these three sectors, benefits from reductions in pollution from utilites account for about half the benefits, and reductions in pollution from industrial point sources and residential- and commercial-fuels combustion account for about 25 percent of the benefits each.

Within the electric utilities, industrial combustion, and residential- and

Table 4–6
Environmental Benefits of Solar Energy under Mid-Range Assumptions

Maximum Practical Case	Maximum Feasible Case
4% of 1978 GNP	7% of 1978 GNP
2.3% of environmental damages	3.9% of environmental damages
$266 (1972 dollars) per capita	$454 (1972 dollars) per capita
$2.75 (1978 dollars) per MM Btu in year 2000	$1.54 (1978 dollars) per MM Btu in year 2000
$4.30 (1972 dollars) per bbl oil equivalent/yr in 1978	$7.38 (1972 dollars) per bbl oil equivalent/yr in 1978

Note: Assumed value of human life is $200,000 (1972 dollars); 2.5 percent is real rate of discount.

Table 4-7
Cumulative Expected Damages, 1975 to 2000: Alternative Scenarios, Alternative Discount Rates (Value of Life = $30,000)
(billions of 1972 dollars)

	Discount Rate = 0.0%			Discount Rate = 2.5%			Discount Rate = 5.0%			Discount Rate = 7.5%			Discount Rate = 10.0%		
	BC	MPC	MFC	BC	MPC	MFC	BC	MPC	MFC	BC	MPC	MFC	BC	MPC	MFC
Electric utilities[a]	399.9	380.4	368.3	288.4	277.0	269.6	216.3	209.5	204.9	168.2	164.0	161.1	135.2	132.5	130.6
Industrial point sources[a]	507.1	497.3	484.5	368.9	363.0	355.1	279.4	275.8	270.8	219.6	217.3	214.1	178.3	176.9	174.7
Residential and commercial fuels combustion	167.8	158.3	157.0	124.1	118.5	117.4	95.0	91.6	90.8	75.0	72.9	72.3	61.0	59.6	59.2
Transportation	176.7	175.4	173.7	127.3	126.5	125.4	95.5	95.0	94.3	74.4	74.1	73.6	60.0	59.7	59.4
Municipal waste water	34.9	34.9	35.2	25.3	25.3	25.5	19.0	19.0	19.2	14.9	14.9	15.0	12.0	12.0	12.1
Urban runoff	112.6	111.9	110.9	86.5	86.1	85.4	69.0	68.7	68.3	56.8	56.6	56.4	48.2	48.0	47.8
Agriculture	41.2	40.9	40.5	30.2	30.1	29.8	23.1	23.0	22.8	18.3	18.8	18.1	14.9	14.8	14.8
Mining and other non-point sources[b]	33.6	33.4	33.0	25.3	25.2	24.9	19.8	19.7	19.5	16.0	15.9	15.8	13.3	13.3	13.2
Total	1473.8	1432.5	1403.0	1076.0	1051.4	1033.1	817.2	802.2	790.5	643.3	634.0	626.4	522.9	517.0	511.9

Note: Value of life is $30,000.

BC = base case; MPC = maximum-practical case; MFC = maximum-feasible case.

[a] Includes both air and water pollution damages.

[b] Includes damages from minerals, ore, and coal mining and milling; and damages from nonurban construction and forestry.

Table 4-8
Cumulative Expected Damages, 1975 to 2000: Alternative Scenarios, Alternative Discount Rates (Value of Life = $100,000)
(billions of 1972 dollars)

	Discount Rate = 0.0%			Discount Rate = 2.5%			Discount Rate = 5.0%			Discount Rate = 7.5%			Discount Rate = 10.0%		
	BC	MPC	MFC	BC	MPC	MFC	BC	MPC	MFC	BC	MPC	MFC	BC	MPC	MFC
Electric Utilities[a]	653.6	623.1	603.8	473.0	455.0	443.3	356.0	345.1	337.9	277.7	271.0	266.4	223.7	219.6	216.6
Industrial point sources[a]	866.8	851.3	830.6	631.9	622.6	609.9	480.0	474.0	466.0	377.9	374.3	369.1	307.5	305.3	301.7
Residential and commercial fuels combustion	285.6	270.4	269.7	211.3	202.3	201.5	161.9	156.4	155.6	127.9	124.5	123.8	104.0	101.8	101.2
Transportation	187.3	185.8	184.0	134.6	133.7	132.6	100.9	100.2	99.5	78.4	78.0	77.5	63.0	62.8	62.4
Municipal waste water	46.7	46.8	47.1	33.9	33.9	34.1	25.5	25.5	25.7	20.0	20.0	20.0	16.1	16.1	16.2
Urban runoff	150.9	150.0	148.7	115.9	115.3	114.4	92.4	92.0	91.5	76.1	75.9	75.5	64.5	64.4	64.0
Agriculture	55.2	54.8	54.2	40.5	40.3	39.9	30.9	30.8	30.5	24.5	24.4	24.2	20.0	19.9	19.9
Mining and other non-point sources[b]	45.0	44.7	44.2	33.9	33.7	33.4	26.5	26.4	26.2	21.4	21.4	21.2	17.9	17.8	17.3
Total	2291.3	2226.9	2182.4	1675.1	1636.8	1609.0	1273.7	1250.5	1232.7	1003.8	989.4	977.8	816.7	807.6	799.8

Note: Value of life is $100,000.

BC = base case; MPC = maximum-practical case; MFC = maximum feasible case.

[a] Includes both air and water pollution damages.

[b] Includes damages from minerals, ore, and coal mining and milling; and damages from nonurban construction and forestry.

Table 4-9
Cumulative Expected Damages, 1975 to 2000: Alternative Scenarios, Alternative Discount Rates (Value of Life = $200,000)
(billions of 1972 dollars)

	Discount Rate = 0.0%			Discount Rate = 2.5%			Discount Rate = 5.0%			Discount Rate = 7.5%			Discount Rate = 10.0%		
	BC	MPC	MFC	BC	MPC	MFC	BC	MPC	MFC	BC	MPC	MFC	BC	MPC	MFC
Electric utilities[a]	1017.5	971.1	941.6	737.7	710.4	692.5	556.2	539.7	528.6	434.6	424.5	417.4	350.7	344.4	339.8
Industrial point sources[a]	1380.4	1356.6	1324.5	1007.4	993.2	973.4	765.6	757.0	744.5	603.8	598.4	590.4	492.0	488.5	483.3
Residential and commercial fuels combustion	454.5	431.0	431.3	336.5	322.5	321.9	257.8	249.3	248.4	203.8	198.5	197.6	165.6	162.3	161.5
Transportation	202.4	200.7	198.7	145.1	144.0	142.8	108.3	107.6	106.9	84.0	83.5	83.0	67.3	67.1	66.7
Municipal waste water	63.5	63.5	64.0	46.0	46.0	46.0	34.7	34.5	34.9	27.1	27.1	27.2	21.9	21.9	22.0
Urban runoff	205.0	203.7	201.9	157.4	156.6	155.4	125.5	125.0	124.2	103.5	103.1	102.6	87.7	87.4	87.0
Agriculture	75.0	74.4	63.6	55.0	54.7	54.2	42.0	41.8	41.5	33.2	33.0	32.9	27.1	27.0	26.9
Mining and other non-point sources[b]	61.2	60.8	60.0	46.0	45.8	45.4	36.0	35.8	35.6	29.1	29.0	28.8	24.3	24.2	24.0
Total	3459.4	3361.9	3295.7	2531.1	2473.2	2431.8	1926.1	1890.9	1864.5	1519.0	1497.2	1479.9	1236.6	1222.8	1211.2

Note: Value of life is $200,000.

BC = base case; MPC = maximum-practical case; MFC = maximum-feasible case.

[a]Includes both air and water pollution damages.

[b]Includes damages from minerals, ore, and coal mining and milling; and damages from nonurban construction and forestry.

Table 4-10
Cumulative Expected Damages, 1975 to 2000: Alternative Scenarios, Alternative Discount Rates (Value of Life = $300,000)
(billions of 1972 dollars)

	Discount Rate = 0.0%			Discount Rate = 2.5%			Discount Rate = 5.0%			Discount Rate = 7.5%			Discount Rate = 10.0%		
	BC	MPC	MFC	BC	MPC	MFC	BC	MPC	MFC	BC	MPC	MFC	BC	MPC	MFC
Electric utilities[a]	1378.9	1316.8	1277.1	1000.7	964.1	940.0	755.0	733.0	718.1	590.5	577.0	567.5	476.8	468.4	462.2
Industrial point sources[a]	1892.2	1860.4	1816.9	1381.6	1362.6	1335.9	1050.6	1039.0	1022.2	829.0	821.7	810.9	675.8	671.2	664.1
Residential and commercial fuels combustion	622.4	590.8	591.9	460.8	442.0	441.6	353.1	341.6	340.7	279.2	272.0	271.0	226.9	222.4	221.4
Transportation	217.5	215.6	213.4	155.5	154.3	153.0	115.8	115.0	114.2	89.5	89.1	88.5	71.7	71.4	71.0
Municipal waste water	80.2	80.2	80.9	58.1	58.1	58.5	43.8	43.8	44.1	34.3	34.3	34.4	27.7	27.7	27.8
Urban runoff	259.1	257.5	255.2	198.9	197.9	196.4	158.6	158.0	157.0	130.7	130.3	129.6	110.8	110.5	110.0
Agriculture	94.8	94.1	93.1	69.6	69.1	68.5	53.1	52.8	52.4	42.0	41.8	41.5	34.3	34.6	34.0
Mining and other non-point sources[b]	77.3	76.8	75.9	58.2	57.9	57.3	45.5	45.3	44.9	36.8	36.7	36.4	30.6	30.6	30.4
Total	4622.3	4492.1	4404.4	3383.4	3306.0	3251.2	2575.6	2528.6	2493.5	2031.9	2002.8	1979.8	1654.6	1636.2	1620.8

Note: Value of life is $300,000.
BC = base case; MPC = maximum-practical case; MFC = maximum-feasible case.
[a] Includes both air and water pollution damages.
[b] Includes damages from minerals, ore, and coal mining and milling; and damages from nonurban construction and forestry.

Table 4-11
Sectors Causing Changes in Net Discounted Environmental Damages
(*billions of 1972 dollars*)

	BC (1H)	MPC (2H)	MFC (3H)	(2H − 1H)	(3H − 1H)	$\frac{2H-1H}{1H} \times 100$	$\frac{3H-1H}{1H} \times 100$
Electric utilities	737.7	710.4	692.5	−27.3(47.2)[a]	−45.2(45.3)	−3.7%	−6.1%
Industrial point sources	1007.4	993.2	973.4	−14.2(24.5)	−34.0(34.1)	−1.4%	−3.4%
Residential and commercial fuel combustion	336.5	322.5	321.9	−14.0(24.2)	−14.6(14.6)	−4.2%	−4.3%
Transportation	145.1	144.0	142.8	−1.1(1.9)	−2.3(2.3)	−.8%	−1.6%
Municipal waste water	46.0	46.0	46.0	0(0)	0(0)	0%	0%
Urban runoff	157.4	156.6	155.4	−.8(1.4)	−2.0(2)	−.5%	−1.3%
Agriculture	55.0	54.7	54.2	−.3(.5)	−.8(.8)	−.5%	−1.4%
Mining and other non-point sources	46.0	45.8	45.2	−.2(.3)	−.8(.8)	−.4%	−1.7%

Note: Value of life is $200,000; discount rate is 2.5 percent.
[a] Numbers in parentheses are percentages of total net benefits from each sector.

Table 4-12
Cumulative Expected Damages and Benefits, 1975 to 2000: Alternative Scenarios by Pollutant
(billions of 1972 dollars)

	BC (1H)	MPC (2H)	MFC (3H)	(1H – 2H)	(1H – 3H)	$\frac{1H - 2H}{1H} \times 100$	$\frac{1H - 3H}{1H} \times 100$
	Damages	Damages	Damages	Benefits	Benefits	% Change, Benefits	% Change, Benefits
Electric utilities							
Part. matter	75.6	74.4	73.3	1.2(1.2)[a]	2.3(1.4)[a]	1.6	3.0
Sulfur oxides	718.0	685.7	667.7	32.3(33.5)	50.3(31.2)	4.5	7.0
Nitrogen oxides	37.0	34.2	33.9	2.8(2.9)	3.1(1.9)	7.6	8.4
Hydrocarbons	0.1	0.1	0.1	0(0)	0(0)	0	0
Carbon monoxide	0.006	0.005	0.005	.001(.001)	0(0)	16.7	0
Water pollution[b]	186.7	176.7	166.5	10.0(10.4)	20.2(12.5)	5.6	10.9
Industry							
Part. matter	608.8	612.3	613.3	-3.4(-3.5)	-4.4(-2.7)	-.6	-.7
Sulfur oxides	506.7	484.0	455.7	22.7(23.5)	5.1(31.6)	4.5	10.0
Nitrogen oxides	35.4	34.0	32.2	1.4(1.5)	3.2(2.0)	4.0	9.0
Hydrocarbons	20.0	19.7	19.3	.3(.3)	.7(.4)	1.5	3.5
Carbon Monoxide	1.1	1.1	1.1	0(0)	0(0)	0	0
Water pollution	208.2	205.7	202.9	2.5(2.6)	5.3(3.3)	1.2	2.6
Residential and commercial							
Part. matter	114.8	121.1	132.7	-6.3(-6.5)	-17.9(-11.1)	-5.5	-15.6
Sulfur oxides	305.1	277.4	267.1	27.7(28.7)	38.0(23.6)	9.1	12.5
Nitrogen oxides	23.2	21.5	20.6	1.7(1.8)	2.6(1.6)	7.3	11.2
Hydrocarbons	11.3	11.1	10.9	.2(.2)	.4(.3)	1.8	3.6
Carbon monoxide	0.03	0.03	0.02	0(0)	.01(0)	0	33.3
Water pollution[c]	63.5	63.5	64.0	0(0)	-.5(-.3)	0	.8
Total	3323.2	3226.7	3162.0	96.5	161.2	2.9	4.9

Note: Value of life is $200,000; discount rate is 0.0 percent.

BC = base case; MPC = maximum-practical case; MFC = maximum-feasible case.

[a] Numbers in parentheses are percentages of net benefits from each sector and pollutant.

[b] Includes both air and water pollution damages.

[c] Includes damages from minerals, ore, and coal mining and milling; and damages from nonurban construction and forestry.

commercial-fuels combustion sectors, it is of interest to know which of the pollutants are responsible for environmental damages and which pollutant reductions are responsible for the environmental benefits of the solar-energy scenarios. Table 4-12 shows this information using a $200,000 value of human life and no discounting (discounting has no significant effect on the distribution of benefits by sector and pollutants). In examining table 4-12, one notices immediately that for all three important sectors by far the largest net environmental benefits from solar energy come from reductions in sulfur oxides. In comparisons of both the maximum-practical and maximum-feasible cases, total benefits from reductions in sulfur oxides are about 86 percent of all environmental benefits. Improvements in water quality in the utility sector account for the bulk of the remainder. Particulates are the only pollutants for which the change from the base case to the high solar scenarios yields disbenefits. As table 4-12 indicates, these disbenefits are due to increases in particulate emissions primarily from the stone- and clay-products industry. These increases result from the expansion in this industry required by solar manufacturing. The residential and commercial disbenefits are due to particulate increases resulting directly from the increased use of wood stoves. The sectors responsible for the increase in particulate damages can be seen in detail in table 4-2.

The environmental benefits from solar energy can be further disaggregated over time and by region. There is little point in disaggregating by time per se because the present value of net benefits is the appropriate economic indicator for use in present-day policy analysis. For general interest, however, time-dependent damages are displayed in table 4-13, in which the assumed value of human life is $200,000.

Tables 4-14 and 4-15 show regional air and water damages in the year 2000 for the base, maximum-practical, and maximum-feasible cases assuming a $200,000 (1972 dollars) value of human life. Columns 4, 5, and 6 of these tables compare the regional damages in each of the three scenarios. The comparison shows clearly that the net benefits from deployment of solar energy in the year 2000 are greatest on the East and West coasts (regions I, II, and IX—see figure 4-1). However, all regions experience some benefits, an important projection from a political point of view.

4.2 Stand-Alone Analysis [a]

4.2.1 Introduction and Summary

In addition to the scenario-dependent emissions presented and analyzed in section 4.1, emissions estimates for individual technologies were generated

---

[a] Kathryn Lawrence wrote the majority of this section.

Table 4–13
National Expected Air Pollution Damages for Alternative Scenarios
(*billions of 1972 dollars*)

	BC	MPC	MFC
	1975		
Electric utilities	23.3	23.3	23.3
Industrial point sources	37.2	37.2	37.2
Residential and commercial fuels			
burning	11.1	11.1	11.1
Transportation	5.0	5.0	5.0
Total	76.6	76.6	76.6
	1985		
Electric utilities	33.0	33.0	32.8
Industrial point sources	43.1	43.1	43.0
Residential and commercial fuels			
burning	20.8	20.7	20.6
Transportation	7.3	7.3	7.3
Total	104.2	104.1	103.7
	1990		
Electric utilities	35.7	35.4	35.0
Industrial point sources	47.4	47.0	45.7
Residential and commercial fuels			
burning	22.8	22.4	21.4
Transportation	7.8	7.8	7.7
Total	113.7	112.6	109.8
	2000		
Electric utilities	34.0	27.8	25.1
Industrial point sources	52.5	49.2	45.7
Residential and commercial fuels			
burning	12.1	8.5	9.9
Transportation	11.2	10.9	10.7
Total	109.8	96.4	91.4

Note: Value of life is $200,000.

BC = base case; MPC = Maximum-practical case; MFC = maximum-feasible case.

by the SEAS model. These runs were termed *stand-alone analyses.* Residual emissions for the energy technologies within SEAS were estimated for each of the snapshot years: 1975, 1985, and 2000. Estimates were developed for three phases of the life cycle: construction of the plant, indirect emissions resulting from plant operation and maintenance (O and M), and direct

Table 4–14
Comparison of Air Damages by Federal Region: Year 2000
(1972 dollars)

Region		1H	2H	3H	Ratios 2H/1H	3H/1H	3H/2H
New England	I	0.395512E + 10	0.338683E + 10	0.240490E + 10	0.86	0.61	0.71
N.Y./N.J.	II	0.307773E + 11	0.222619E + 11	0.219183E + 11	0.72	0.71	0.98
Middle Atlantic	III	0.115800E + 11	0.112127E + 11	0.106725E + 11	0.97	0.92	0.95
Southeast	IV	0.809066E + 10	0.775392E + 10	0.744187E + 10	0.96	0.92	0.96
Great Lakes	V	0.304631E + 11	0.289484E + 11	0.274663E + 11	0.95	0.90	0.95
South Central	VI	0.701462E + 10	0.670917E + 10	0.659762E + 10	0.96	0.94	0.98
Central	VII	0.226650E + 10	0.222513E + 10	0.218900E + 10	0.98	0.97	0.98
Mountain	VIII	0.152516E + 10	0.136760E + 10	0.125151E + 10	0.90	0.82	0.92
West	IX	0.119971E + 11	0.105077E + 11	0.949766E + 10	0.88	0.79	0.90
Northwest	X	0.213068E + 10	0.202671E + 10	0.196029E + 10	0.95	0.92	0.97
National total		0.109800E + 12	0.963999E + 11	0.913997E + 11	0.88	0.83	0.95

Note: Value of life is $200,000.
1H = base case (high environmental control).
2H = Maximum-practical case (high environmental control).
3H = Maximum-feasible case (high environmental control).

Table 4-15
Comparison of Water Damages by Federal Region: Year 2000
(1972 dollars)

Region		1H	2H	3H	Ratios		
					2H/1H	3H/1H	3H/2H
New England	I	0.124275E + 10	0.125099E + 10	0.110330E + 10	1.01	0.89	0.88
N.Y./N.J.	II	0.357087E + 10	0.342385E + 10	0.332806E + 10	0.96	0.93	0.97
Middle Atlantic	III	0.380230E + 10	0.363700E + 10	0.336481E + 10	0.96	0.88	0.93
Southeast	IV	0.811636E + 10	0.767014E + 10	0.694435E + 10	0.95	0.86	0.91
Great Lakes	V	0.947530E + 10	0.915423E + 10	0.870321E + 10	0.97	0.92	0.95
South Central	VI	0.791608E + 10	0.712484E + 10	0.702527E + 10	0.90	0.89	0.99
Central	VII	0.107717E + 10	0.102911E + 10	0.986439E + 09	0.96	0.92	0.96
Mountain	VIII	0.529874E + 09	0.515249E + 09	0.497296E + 09	0.97	0.94	0.97
West	IX	0.222537E + 10	0.206473E + 10	0.194129E + 10	0.93	0.87	0.94
Northwest	X	0.343914E + 09	0.329855E + 09	0.305938E + 09	0.96	0.89	0.93
National total		0.383000E + 11	0.362000E + 11	0.342000E + 11	0.95	0.89	0.94

Note: Value of life is $200,000.

1H = Base case (high environmental control).

2H = Maximum-practical case (high environmental control).

3H = Maximum-feasible case (high environmental control).

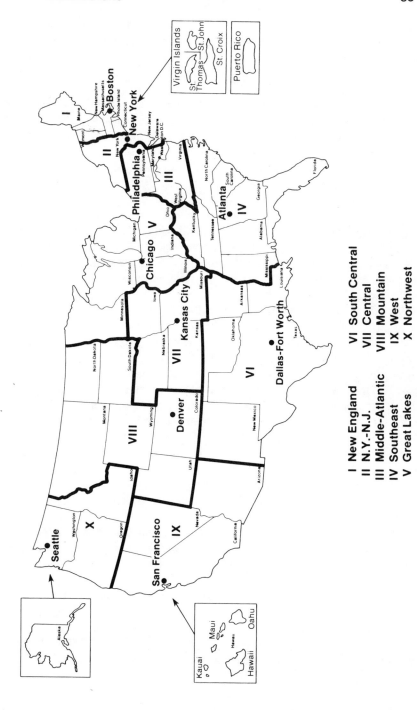

Figure 4-1. Federal Regions of the United States

I New England VI South Central
II N.Y.-N.J. VII Central
III Middle-Atlantic VIII Mountain
IV Southeast IX West
V Great Lakes X Northwest

emissions associated with plant operation. Only seven residual categories are reported for the stand-alone analyses: five air emissions (particulates, sulfur oxides [SO_x], nitrogen oxides [NO_x], carbon monoxide [CO], and hydrocarbons) and two water pollutants (biological oxygen demand [BOD] and suspended solids).

The emissions associated with the solar-energy technologies tended to be *front end,* that is, occurring primarily during the construction phase. The solar facilities' construction emissions estimated with the SEAS model generally exceeded those for conventional energy facilities. By contrast, the direct operation and maintenance emissions from conventional energy facilities exceeded analogous solar-energy facility phases, particularly when the entire fuel cycle was considered. To derive an estimate of the long-term emissions of each energy-facility option, all construction and direct and indirect operation emissions were summed and adjusted for conversion efficiencies throughout the life-cycle of an energy facility delivering 10^{12} Btu/yr. The entire fuel and energy distribution cycles were considered, and emissions were summed for each facility's lifetime (for example, thirty years for a coal steam-electric facility, fifty years for wood stoves, and so on). Results of these calculations indicate that, in the long run, cumulative emissions of particulates, SO_x, and NO_x from coal steam-electric facilities meeting new source performance standards generally exceed those of solar-energy facilities (with the exception of uncontrolled particulate releases from wood stoves). By contrast, cumulative lifetime emissions of hydrocarbons, CO, BOD, and suspended solids associated with solar facilities often exceeded those associated with coal facilities. These levels were due almost entirely to materials processing and plant construction. In addition, the life-cycle emissions of particulates, SO_x, and CO associated with the centralized photovoltaic-energy option were generally greater than those associated with the distributed photovoltaic deployment mode. The life-cycle emissions of the more material-intensive-per-unit-energy-output solar-energy systems were relatively high. For example, emissions associated with flat-plate AIPH facilities were slightly larger than for parabolic-trough and parabolic-dish facilities; with the exception of particulate and carbon-monoxide releases, SHACOB active facility-derived emissions exceeded the passive option; and the life-cycle emissions of the solar-thermal-central-receiver option exceeded those of wind energy conversion facilities.

Presented in the following sections are: (1) results of stand-alone SEAS runs and assumptions underlying the runs; (2) a discussion of the residuals produced by solar-energy systems as compared to conventional energy systems; and (3) a summary of significant environmental implications not accounted for within the SEAS model for the solar-energy systems.

4.2.2 SEAS Stand-Alone Analysis Results: 1985 and 2000

To assess the relative environmental implications of individual energy technologies and perform a partial check of residuals projections from the scenario analyses, the SEAS model was programmed to calculate residuals per 10^{12} Btu delivered energy from each of the energy technologies contained within the model. As previously noted, calculations were made for three phases of the life cycle: construction, indirect O and M, and direct O and M. The calculations for the solar technologies do not include any conventional energy backup that may be required.

The construction phase is defined to include mining raw materials, material processing, fabrication of system components, plant construction or system installation, plus emissions from materials transport and construction vehicles. Because the SEAS model employs a 200-by-200 input-output matrix that is representative of the entire economy for the estimation procedure, residual estimates also include emissions released from manufacture of equipment to process the raw materials (for example, a steel foundry), the emissions from producing mining equipment, and so forth.

Construction impacts are based on the assumption that enough systems are built within the snapshot year to provide 10^{12} Btu/yr delivered energy. Although actual construction activities occur over several years, all construction emissions are specified to occur within the first year in this portion of the model. *Lifetime residuals* are defined as the sum of emissions resulting from construction of all the plants required to deliver 10^{12} Btu/yr.

Annual construction emissions were not provided by the SEAS model. Lifetime residuals were divided by plant lifetime (in years) to approximate emissions under a steady-state condition. The *steady-state condition* was defined as follows: at the beginning of the year, enough plants are operating to provide 10^{12} Btu/yr. Plants are assumed to be of various ages. Within any year, some plants' useful lifetimes will expire and, therefore, they must be replaced to maintain an output of 10^{12} Btu/yr. The replacement frequency for the entire complex of a plant type is assumed to be 1/(plant lifetime). Alternatively, the probability that a particular plant will be retired in a given year is 1/(plant lifetime). This replacement factor was applied to the lifetime emissions to provide an estimate of annual emissions.

The second life-cycle phase includes all secondary activities associated with plant O and M. Estimates include residuals generated for processing the materials required to replace damaged, malfunctioning, or spent system components on an annual basis; for example, replacement of broken heliostats, photovoltaic panels, and working fluids. Because an infinite chain approach is utilized within the model, emissions released from combustion of fuel required for material processing or from processing equipment manufacture are also included within the indirect-O and M- residuals tabulation.

Estimates of indirect O and M residuals within the stand-alone analyses are more comprehensive for the solar than for the conventional energy technologies. Residuals from operation and maintenance of conventional energy technologies (for example, coal) are not accounted for within the stand-alone analysis.[3] Emissions associated with acquisition and processing of conventional fuels are reported as separate cycles. These emissions are not reported in tables 4–23 to 4–29. However, this is not the case for the nonconventional energy technologies that require fuel inputs other than direct solar radiation. For example, residuals emitted from silviculture farms and agriculture and forest residue collection are estimated. Parallel operations (coal mining and processing, oil drilling and refining, and so forth) for the conventional technologies are not accounted for in the indirect O and M estimates of a coal steam-electric facility of oil-fired power plant. Thus, at first inspection, the indirect emissions attributable to solar-energy technologies far exceed those attributable to conventional energy technologies. This is not always the case when the emissions associated with fuel acquisition and processing of conventional fuels are included in indirect O and M-residuals estimates (see section 4.2.3 for this tabulation).

The final phase for which residuals were estimated is plant operation. Plant-operation residuals are those attributable solely to the energy conversion process. Representative residuals are particulates released from coal, wood, or residue combustion. For all solar energy technologies other than biomass, the fuel converted during plant operation is radiant energy. As a result, most of the solar energy technologies emit no direct operating residuals.

SEAS-generated estimates of residuals emitted by the solar-energy technologies in 1985, and 2000 are shown in tables 4–16 through 4–22, and by the nonsolar energy technologies in tables 4–23 through 4–29. As illustrated in the tables, annual residuals for each life-cycle phase decrease with time. These declines are due to changes in input assumptions entered into the SEAS model. A brief summary of input assumptions influencing results of the SEAS stand-alone projections is listed in table 4–30. Since the input assumptions used in the stand-alone analysis are identical to those used for the scenario analysis, identical caveats apply. In particular, since stand-alone results are ultimately based on the cost data for each solar and conventional technology, any error in judgment here will propagate through the model (essentially linearly) to the results.

The input assumption most strongly affecting residual projections is environmental-control regulations. Compliance with Environmental Protection Agency and/or state emission standards (whichever were the more stringent) was assumed to be a phased process completed in 1983.

Table 4–16
Annual Emissions of Particulates from Solar Energy Facilities
(SEAS Model)
(*tons/10¹² Btu*)

Technology	Construction 1985	Construction 2000	Indirect O & M 1985	Indirect O & M 2000	Direct O & M 1985	Direct O & M 2000	Totals 1985	Totals 2000
SHACOB								
Passive	144	142	0	0	0	0	144	142
Active	69	67	21	20	0	0	90	87
AIPH								
Flat-plate	22	20	29	28	0	0	51	48
Parabolic trough	20	21	24	23	0	0	44	44
Parabolic dish	8	8	34	34	0	0	42	42
Solar thermal	44	30	44	43	0	0	88	73
Photovoltaics								
Distributed	114	61	17	7	0	0	131	74
Centralized	109	65	53	29	0	0	162	94
WECS	14	10	10	7	0	0	24	17
Biomass								
Residue collection	neg[a]	neg	9	9	neg	neg	10	10
Silvicultural farms	1	1	37	36	2	2	40	39
Anaerobic digestion	3	3	33	32	0	0	35	35
Pyrolysis	1	neg	20	19	2	2	23	21
Steam electric	17	16	102	99	56	56	175	171
Wood stoves	neg	neg	0	0	5091	5092	5092	5093
Municipal wastes	4	3	15	15	0	0	19	18

[a] neg. = negligible, less than 1.

Table 4–17
Annual Emissions of Sulfur Oxides from Solar-Energy Facilities
(SEAS Model)
(*tons/10¹² Btu*)

Technology	Construction 1985	Construction 2000	Indirect O & M 1985	Indirect O & M 2000	Direct O & M 1985	Direct O & M 2000	Totals 1985	Totals 2000
SHACOB	50	42	0	0	0	0	50	42
Passive								
Active	49	43	60	52	0	0	109	95
AIPH								
Flat plate	34	32	21	18	0	0	55	50
Parabolic trough	28	35	23	20	0	0	51	55
Parabolic dish	20	18	14	12	0	0	34	30
Solar thermal	90	70	33	29	0	0	123	99

Table 4–17 continued

Technology	Construction		Indirect O & M		Direct O & M		Totals	
	1985	2000	1985	2000	1985	2000	1985	2000
Photovoltaics								
Distributed	604	366	32	11	0	0	636	377
Centralized	579	393	62	29	0	0	641	422
WECS	26	24	14	8	0	0	40	32
Biomass								
Residue collection	1	1	19	16	—[a]	< 1	20	18
Silvicultural farms	3	2	69	60	4	4	76	66
Anaerobic digestion	96	8	43	36	90	90	229	134
Pyrolysis	2	1	56	53	9	9	67	63
Steam electric	22	19	106	96	28	28	156	143
Wood stoves	2	1	0	0	0	0	2	1
Municipal wastes	8	7	53	47	0	0	61	54

[a] Line indicates technology not operational in 1985; therefore, direct O & M residuals = 0.

Table 4–18
Annual Emissions of Nitrogen Oxides from Solar-Energy Facilities (SEAS Model)
(tons/10^{12} Btu)

Technology	Construction		Indirect O & M		Direct O & M		Totals	
	1985	2000	1985	2000	1985	2000	1985	2000
SHACOB								
Passive	60	47	0	0	0	0	60	70
Active	36	30	45	39	0	0	81	69
AIPH								
Flat plate	24	21	18	14	0	0	42	35
Parabolic trough	24	25	18	15	0	0	42	40
Parabolic dish	22	17	10	9	0	0	32	26
Solar thermal	82	60	20	17	0	0	102	77
Photovoltaics								
Distributed	512	260	25	8	0	0	537	268
Centralized	366	222	40	19	0	0	406	241
WECS	20	17	12	6	0	0	32	23
Biomass								
Residue collection	1	1	26	17	—[a]	< 6	27	24
Silvicultural farms	3	2	87	78	33	32	123	112
Anaerobic digestion	9	8	37	30	80	80	126	118
Pyrolysis	2	1	45	37	10	10	57	48
Steam electric	20	16	192	117	314	314	526	447
Wood stoves	1	1	0	0	0	0	1	1
Municipal wastes	8	6	30	28	0	0	38	34

[a] Line indicates technology not operational in 1985; therefore, direct O & M residuals = 0.

Table 4–19
Annual Emissions of Carbon Monoxide from Solar-Energy Facilities
(SEAS Model)
(*tons/10¹² Btu*)

Technology	Construction 1985	Construction 2000	Indirect O & M 1985	Indirect O & M 2000	Direct O & M 1985	Direct O & M 2000	Totals 1985	Totals 2000
SHACOB								
Passive	163	78	0	0	0	0	163	78
Active	93	48	48	32	0	0	141	80
AIPH								
Flat plate	54	29	31	32	0	0	85	61
Parabolic trough	74	42	24	17	0	0	98	59
Parabolic dish	121	89	19	15	0	0	140	104
Solar thermal	212	123	74	68	0	0	286	191
Photovoltaics								
Distributed	1077	260	39	11	0	0	1116	271
Centralized	1022	385	79	37	0	0	1101	422
WECS	61	27	42	23	0	0	103	50
Biomass								
Residue collection	2	1	139	68	— [a]	< 8	141	77
Silvicultural farms	13	7	140	88	44	44	197	139
Anaerobic digestion	22	14	74	50	0	0	96	64
Pyrolysis	5	3	112	56	5	5	122	64
Steam electric	45	28	1196	545	112	112	1353	685
Wood stoves	4	2	0	0	0	0	4	2
Municipal wastes	29	16	58	50	0	0	87	66

[a] Line indicates technology not operational in 1985; therefore, direct O & M residuals = 0.

Table 4–20
Annual Emissions of Hydrocarbons from Solar-Energy Facilities
(SEAS Model)
(*tons 10¹² Btu*)

Technology	Construction 1985	Construction 2000	Indirect O & M 1985	Indirect O & M 2000	Direct O & M 1985	Direct O & M 2000	Totals 1985	Totals 2000
SHACOB								
Passive	15	11	0	0	0	0	15	11
Active	8	5	12	11	0	0	20	16
AIPH								
Flat plate	5	3	6	5	0	0	11	8
Parabolic trough	6	3	4	3	0	0	10	6
Parabolic dish	9	7	4	4	0	0	13	11
Solar thermal	22	13	22	18	0	0	44	31

Table 4–20 continued

Technology	Construction		Indirect O & M		Direct O & M		Totals	
	1985	2000	1985	2000	1985	2000	1985	2000
Photovoltaics								
Distributed	180	77	7	3	0	0	187	80
Centralized	97	37	24	11	0	0	121	48
WECS	4	2	4	2	0	0	8	4
Biomass								
Residue collection	neg[a]	neg	11	7	0	1	11	8
Silvicultural farms	1	1	23	18	7	7	31	26
Anaerobic digestion	2	1	12	10	0	0	14	11
Pyrolysis	neg	neg	14	10	1	1	15	11
Steam electric	5	4	124	85	373	373	502	462
Wood stoves	neg	neg	0	0	0	0	neg	neg
Municipal wastes	3	2	5	4	0	0	8	6

[a] neg = negligible, less than 1.

Table 4–21
Annual Water Emissions in Biological Oxygen Demand from Solar-Energy Facilities (SEAS Model)
(tons 10^{12} Btu)

Technology	Construction		Indirect O & M		Direct O & M		Totals	
	1985	2000	1985	2000	1985	2000	1985	2000
SHACOB								
Passive	neg[a]	neg	0	0	0	0	neg	neg
Active	neg	neg	neg	neg	0	0	1	1
AIPH								
Flat plate	neg	neg	neg	neg	0	0	neg	neg
Parabolic trough	neg	neg	neg	neg	0	0	neg	neg
Parabolic dish	neg	neg	neg	neg	0	0	neg	neg
Solar thermal	1	neg	1	neg	0	0	2	1
Photovoltaics								
Distributed	10	4	neg	neg	0	0	10	4
Centralized	3	2	2	1	0	0	5	3
WECS	neg	neg	neg	neg	0	0	neg	neg
Biomass								
Residue collection	neg	neg	neg	neg	0	0	neg	neg
Silvicultural farms	neg	neg	1	1	0	0	1	1
Anaerobic digestion	neg	neg	neg	neg	0	0	neg	neg
Pyrolysis	neg	neg	neg	neg	0	0	neg	neg
Steam electric	neg	neg	2	1	0	0	2	1
Wood stoves	neg	neg	0	0	0	0	neg	neg
Municipal wastes	neg	neg	neg	neg	0	0	neg	neg

[a] neg = negligible, less than 1.

Table 4–22
Annual Water Emissions of Suspended Solids from Solar-Energy Facilities (SEAS Model)
(*tons /10¹² Btu*)

Technology	Construction		Indirect O & M		Direct O & M		Totals	
	1985	*2000*	*1985*	*2000*	*1985*	*2000*	*1985*	*2000*
SHACOB								
Passive	neg[a]	neg	0	0	0	0	neg	neg
Active	neg	neg	1	1	0	0	1	1
AIPH								
Flat plate	neg	neg	neg	neg	0	0	neg	neg
Parabolic trough	neg	neg	neg	neg	0	0	neg	neg
Parabolic dish	neg	neg	neg	neg	0	0	neg	neg
Solar thermal	1	neg	1	1	0	0	2	1
Photovoltaics								
Distributed	9	4	neg	neg	0	0	9	4
Centralized	4	2	2	1	0	0	6	3
WESC	neg	neg	neg	neg	0	0	neg	neg
Biomass								
Residue collection	neg	neg	neg	neg	0	0	neg	neg
Silvicultural farms	neg	neg	1	1	0	0	1	1
Anaerobic digestion	neg	neg	neg	neg	0	0	neg	neg
Pyrolysis	neg	neg	neg	neg	0	0	neg	neg
Steam electric	neg	neg	2	2	0	0	2	2
Wood stoves	neg	neg	0	0	0	0	neg	neg
Municipal wastes	neg	neg	neg	neg	0	0	neg	neg

[a] neg = negligible, less than 1.

Table 4–23
Annual Emissions of Particulates from Conventional Energy Facilities (SEAS Model)
(*tons/10¹² Btu*)

Technology	Construction		Indirect O & M [a]		Direct O & M		Totals	
	1985	*2000*	*1985*	*2000*	*1985*	*2000*	*1985*	*2000*
Coal steam electric	7	6	—[b]	—	262	151	269	157
Oil steam electric	6	5	—	—	60	65	66	70
Gas steam electric	5	4	—	—	14	14	19	18
Synfuel coal	2	1	—	—	8	5	10	6
Synfuel, low Btu gas	1	1	—	—	6	7	7	8
Nuclear steam electric	5	4	—	—	0	0	5	4
Onshore gas	2	2	—	—	0	0	2	2

[a] Impacts associated with fuel acquisition and processing are presented in section 4.2.3.
[b] A line indicates that data were not estimated.

Table 4–24

Annual Emissions of Sulfur Oxides from Conventional Energy Facilities (SEAS Model)

(*tons/10^{12} Btu*)

Technology	Construction		Indirect O & M [a]		Direct O & M		Totals	
	1985	2000	1985	2000	1985	2000	1985	2000
Coal steam electric	15	13	—[b]	—	3359	1839	3374	1852
Oil steam electric	14	12	—	—	1410	1372	1424	1384
Gas steam electric	11	10	—	—	9	9	20	19
Synfuel, coal	4	3	—	—	12	16	16	19
Synfuel, low Btu gas	2	2	—	—	99	120	101	122
Nuclear steam electric	10	9	—	—·	0	0	10	9
Onshore gas	4	3	—	—	neg[c]	neg	4	3

[a] Impacts associated with fuel acquisition and processing are presented in section 4.2.3.

[b] A line indicates that data were not estimated.

[c] neg = negligible, less than 1.

Table 4–25

Annual Emissions of Nitrogen Oxides from Conventional Energy Facilities (SEAS Model)

(*tons/10^{12} Btu*)

Technology	Construction		Indirect O & M [a]		Direct O & M		Totals	
	1985	2000	1985	2000	1985	2000	1985	2000
Coal steam electric	13	11	—[b]	—	1248	1024	1261	1035
Oil steam electric	12	10	—	—	848	773	860	783
Gas steam electric	10	8	—	—	1050	1050	1060	1058
Synfuel, coal	3	3	—	—	37	34	40	37
Synfuel, low Btu gas	2	1	—	—	61	69	63	70
Nuclear steam electric	9	8	—	—	0	0	9	8
Onshore gas	3	3	—	—	5	5	8	8

[a] Impacts associated with fuel acquisition and processing are presented in section 4.2.3.

[b] A line indicates that data were not estimated.

Table 4–26
Annual Emissions of Carbon Monoxide from Conventional Energy Facilities (SEAS Model)
(*tons/10¹² Btu*)

Technology	Construction		Indirect O & M [a]		Direct O & M		Totals	
	1985	2000	1985	2000	1985	2000	1985	2000
Coal steam electric	31	21	—[b]	—	65	62	96	83
Oil steam electric	29	20	—	—	44	42	73	62
Gas steam electric	23	16	—	—	26	26	49	42
Synfuel, coal	9	7	—	—	0	0	9	7
Synfuel, low Btu gas	5	3	—	—	5	5	10	8
Nuclear steam electric	22	15	—	—	0	0	22	15
Onshore gas	12	10	—	—	2	2	14	12

[a] Impacts associated with fuel acquisition and processing are presented in section 4.2.3.
[b] A line indicates that data were not estimated.

Table 4–27
Annual Emissions of Hydrocarbons from Conventional Energy Facilities (SEAS Model)
(*tons/10¹² Btu*)

Technology	Construction		Indirect O & M [a]		Direct O & M		Totals	
	1985	2000	1985	2000	1985	2000	1985	2000
Coal steam electric	3	2	—[b]	—	19	19	22	21
Oil steam electric	3	2	—	—	14	16	17	18
Gas steam electric	2	1	—	—	2	2	4	3
Synfuel, coal	1	1	—	—	1	1	2	2
Synfuel, low Btu gas	neg[c]	neg	—	—	1	2	1	2
Nuclear steam electric	2	1	—	—	0	0	2	1
Onshore gas	1	1	—	—	neg	neg	1	1

[a] Impacts associated with fuel acquisition and processing are presented in section 4.2.3.
[b] A line indicates that data were not estimated.
[c] neg = negligible, less than 1.

Table 4–28

Annual Water Emissions in Biological Oxygen Demand from Conventional Energy Facilities (SEAS Model)

(*tons/10^{12} Btu*)

Technology	Construction		Indirect O & M[a]		Direct O & M		Totals	
	1985	2000	1985	2000	1985	2000	1985	2000
Coal steam electric	neg[b]	neg	—[c]	—	1	1	1	1
Oil steam electric	neg	neg	—	—	2	2	2	2
Gas steam electric	neg	neg	—	—	2	2	2	2
Synfuel, coal	neg	neg	—	—	0	neg	neg	neg
Synfuel, low Btu gas	neg	neg	—	—	11	11	11	11
Nuclear steam electric	neg	neg	—	—	0	neg	neg	neg
Onshore gas	neg	neg	—	—	0	0	neg	neg

[a] Impacts associated with fuel acquisition and processing are presented in section 4.2.3.

[b] neg = negligible, less than 1.

[c] A line indicates that data were not estimated.

Table 4–29

Annual Water Emissions of Suspended Solids from Conventional Energy Facilities (SEAS Model)

(*tons/10^{12} Btu*)

Technology	Construction		Indirect O & M[a]		Direct O & M		Totals	
	1985	2000	1985	2000	1985	2000	1985	2000
Coal steam electric	neg[b]	neg	—[c]	—	neg	neg	1	neg
Oil steam electric	neg	neg	—	—	1	1	1	1
Gas steam electric	neg	neg	—	—	1	1	1	1
Synfuel, coal	neg	neg	—	—	0	neg	neg	neg
Synfuel, low Btu gas	neg	neg	—	—	0	0	neg	neg
Nuclear steam electric	neg	neg	—	—	0	neg	neg	neg
Onshore gas	neg	neg	—	—	0	0	0	0

[a] Impacts associated with fuel acquisition and processing are presented in section 4.2.3.

[b] neg = negligible, less than 1.

[c] A line indicates that data were not estimated.

Table 4–30
Modeling Assumptions: SEAS Stand-Alone Residuals Projections

Assumptions	Year		
	1975	*1985*	*2000*
Materials requirements	high quantities of inputs required, especially for nonconventional technologies	little change from 1975	for most technologies, materials inputs have been reduced through technological improvements
Environmental Control regulations	environmental emission standards not fully enforced; BACT[a] not yet applied to all industries	all environmental standards met by 1983; BACT applied	little change from 1985
Economic structure	base-case input assumptions; see section 3.3	little change from 1975	significant changes made with respect to 1975; see section 3.3 for a discussion

[a]BACT = best available control technologies.

Strict emission standards in effect in 1983 will also apply for the projections for the year 2000. Decreases in residuals estimates from 1985 to 2000, therefore, are not attributable to changes in emission standards. Variations and decreases in projections between 1985 and 2000 are due to the other input assumptions shown in table 4–30; that is, changes in materials requirements and national economic structure. Reduction in the quantity of materials necessary for system fabrication and plant construction was assumed to result from technological improvements in the developing energy technologies. The improvements are further assumed to be targeted at reducing costs. Changes in the materials-input assumptions will most acutely affect the solar-energy technologies. Alterations in economic-structure assumptions (discussed in section 2.1.3) will affect all the energy technologies contained within the model.

4.2.3 Comparison of Solar and Conventional
Stand-Alone Energy Facilities

A comparison of the criteria air emissions and water pollutants can provide a useful guide to the effects of displacing one energy option by another. SEAS-generated residuals for the solar-energy technologies present life-cycle emissions (exclusive of residuals generated during plant decommission). However, SEAS estimates of stand-alone residuals for the conventional (nonsolar) technologies are for the plant construction and operation phases. The residuals released during fuel acquisition (for example, coal mining and processing) were not automatically estimated within the SEAS stand-alone analyses. (The SEAS scenario analyses include the indirect residuals associated with the fuel cycle.) However, construction and operating residuals associated with fuel acquisition and processing were obtainable as separate data. The SEAS estimates for releases of particulates, sulfur oxides, and nitrogen oxides for the coal cycle are shown in tables 4-31, 4-32, and 4-33. These estimates represent the indirect O and M residuals for the conventional energy conversion facilities. Although not presented in tabular form in this section, similar data are available for hydrocarbon, carbon monoxide, BOD, and suspended solids releases.

It should be noted again that comparisons of life-cycle emissions are valid only for the selected systems modeled within SEAS and the input assumptions used (see chapter 3). Further, environmental-effects data for many of the solar-energy technologies are not available, and system technologies are undergoing developmental changes that result in different life-cycle emission levels. The data should not be interpreted as strong evidence of the relative environmental effects of individual solar technologies. Rather, the data gives general indication of the relative environmental impact of typical solar-energy systems versus conventional technologies.

In all instances, the residuals associated with constructing solar facilities exceed those associated with conventional facilities. This relationship is due entirely to the material intensiveness of solar technologies relative to conventional technologies. The most favorable comparison of construction residuals occurs between WECS and coal steam-electric facilities, where WECS residuals are larger by a factor of 1.2 to 2.4, depending on the residual.

Indirect O and M residuals associated with the solar facilities are those emissions associated with producing materials to replace spent or broken components; for example, glass for heliostats and working fluids. Indirect O and M residuals for conventional energy facilities are those associated with fuel acquisition and processing, as shown in tables 4-31 to 4-33. In contrast to the construction and indirect O and M phases, the solar-electric facilities modeled in SEAS (with the exception of the biomass systems) emit

Table 4–31
Annual Emissions of Particulates from Coal Acquisition and Delivery
(*tons/10¹² Btu*)

	Phase[a]	Construction[c]		Direct O & M		Total	
	Lifetime[b]	1985	2000	1985	2000	1985	2000
Coal cycle							
Mining, western surface	20	0.13	0.11	0.80	1.06	0.93	1.11
Mining, eastern underground	20	0.21	0.18	0	0	0.21	0.18
Coal preparation plants	20	0.07	0.06	0.80	0.80	0.87	0.86
Electric utilities, transmission lines	40	2.20	2.10	0	0	2.20	2.21
Electric utilities, distribution	40	1.50	1.20	0	0	1.50	1.20

[a] No indirect O & M residuals are emitted.
[b] Years.
[c] Life-cycle emissions divided by lifetime.

Table 4–32
Annual Emissions of Sulfur Oxides from Coal Acquisition and Delivery
(*tons/10¹² Btu*)

	Phase[a]	Construction[c]		Direct O & M		Total	
	Lifetime[b]	1985	2000	1985	2000	1985	2000
Coal cycle							
Mining, western surface	20	0.32	0.29	0.10	0.10	0.42	0.39
Mining, eastern underground	20	0.51	0.46	0	0	0.51	0.46
Coal preparation plants	20	0.16	0.14	0.01	0.01	0.17	0.15
Electric utilities, transmission lines	40	5.40	4.70	0	0	5.40	4.70
Electric utilities, distribution	40	4.50	4.10	0	0	4.50	4.10

[a] No indirect O & M residuals are emitted.
[b] Years.
[c] Life-cycle emissions divided by lifetime.

no residuals during operation of the facilities throughout their twenty to thirty-year lifetimes. Thus they are more attractive to operate than conventional electric energy facilities. However, during operation the biomass steam-electric facilities modeled emit more particulates than any of the conventional facilities; more SO_x, HC, and CO than nuclear and gas-fired steam-electric facilties; and more NO_x than the nuclear facilities for which NO_x emissions during operation are zero.

Table 4–33

Annual Emissions of Nitrogren Oxides from Coal Acquisition and Delivery
(*tons/10^{12} Btu*)

	Phase[a]	Construction[c]		Direct O & M		Total	
	Lifetime[b]	1985	2000	1985	2000	1985	2000
Coal cycle							
Mining, western surface	20	0.30	0.20	1.90	1.70	2.20	1.90
Mining, eastern underground	20	0.44	0.37	0	0	0.44	0.37
Coal preparation plants	20	0.14	0.12	0.60	0.60	0.74	0.72
Electric utilities, transmission lines	40	4.10	3.40	0	0	4.10	3.40
Electric utilities, distribution	40	3.40	2.90	0	0	3.40	2.90

[a] No indirect O & M residuals are emitted.

[b] Years.

[c] Life-cycle emissions divided by lifetime.

    The bulk of the life-cycle emissions for the solar technologies are front end. The SEAS model estimates emissions only to the year 2000, the point where the solar-energy technologies are beginning to make significant market penetration. Thus under the high solar scenarios the model simulates a scale-up phenomenon where the rate of solar facilities construction is rapid (see section 4.1). The long-term, steady-state, potential environmental benefits of operating solar-energy conversion facilities (which often emit no air pollutants) for twenty to thirty years, therefore, are not portrayed by the SEAS estimates.

    The long-term condition can be simulated by calculating the life-cycle emissions for an energy-conversion option. This calculation was performed for thirteen solar-energy technologies and one conventional technology, coal steam-electric. The life-cycle emission estimate for each technology option was determined by summing emissions from (1) fuel processing throughout the system's lifetime; (2) fuel preparation; (3) fuel conversion; and (4) electricity transmission and distribution. All calculations were corrected for conversion efficiencies throughout the cycle to assure that 10^{12} Btu/yr are delivered at the end of the cycle. The results of this process are shown in figures 4–2 through 4–8.

    In general, the solar life-cycle emissions are far below those of the coal steam-electric option. This is particularly true for life-cycle releases of sulfur oxides. It is also interesting to note that life-cycle emissions of particulates, sulfur oxides, and carbon monoxide associated with the centralized cadmium-sulfide photovoltaic-deployment option are higher than those

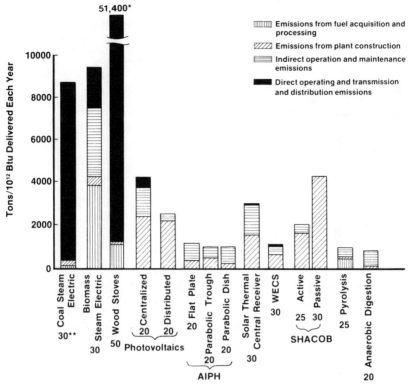

*Emissions are uncontrolled for wood stoves; all other technologies employ best available control technologies and/or meet New Source Performance Standards.

**Plant lifetime in years.

Figure 4–2. Cumulative Life-Cycle Particulate Emissions, 1985 Data

associated with the distributed deployment option that was modeled. In general, the emissions associated with the SHACOB active system exceed those of the passive facilities with the exception of particulate and carbon-monoxide releases. The passive system considered in the study is a Trombe-wall design that is comprised of considerable quantities of concrete. Life-cycle particulate releases for the passive facilities occurs almost exclusively from mining and processing the cement, sand, and gravel required for concrete production. Nonconcrete passive designs would be significantly different from the trombe wall considered in this book. Of the solar-thermal options considered, life-cycle emissions associated with the central-receiver facilities are larger than the AIPH flat-plate and parabolic-trough and the parabolic-dish total-energy facilities.

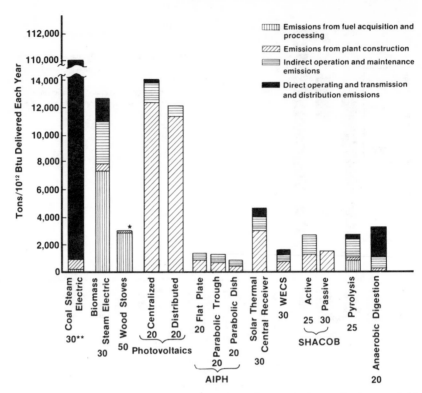

*Emissions are uncontrolled for wood stoves; all other technologies employ best available control technologies and/or meet New Source Performance Standards.
**Plant lifetime in years.

Figure 4–3. Cumulative Life-Cycle Sulfur Oxide Emissions, 1985 Data

4.2.4 Solar-Energy Systems: Other Environmental Issues

Utilization of the SEAS model, in conjunction with the AMBIENT and BENEFITS modules, allows projections of the environmental damage resulting from the five criteria air pollutants plus an array of water pollutants associated with a specific energy-deployment scenario. Because projections of the monetary damages of air pollution are limited to a small number of residuals categories, there are many potential impacts not represented within the damage projections. This is true for both the solar and nonsolar-energy technologies. Some of the impact areas not accounted for include: (1) many of the health effects of coal mining, such as black lung disease; (2) many of the environmental and aesthetic impacts associated

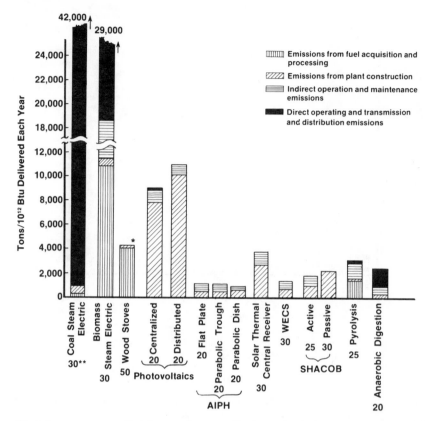

*Emissions are uncontrolled for wood stoves; all other technologies employ best available control technologies and/or meet New Source Performance Standards.

**Plant lifetime in years.

Figure 4–4. Cumulative Life-Cycle Nitrogen Oxide Emissions, 1985 Data

with fuel acquisition; for example, strip mining of coal or a silviculture farm; (3) operating emissions not within the residual categories; (4) other environmental impacts within the life cycle; for example, microclimate alterations, disposal of waste products; and (5) the timing of environmental impacts and residuals emissions within the life cycle. The focus, therefore, is on those environmental effects, both positive and negative, that are not accounted for within the SEAS model. Many of these impacts are not unique to the solar-energy technologies. Fugitive dust, noise, site disturbance, and so forth, will occur during construction of any large industrial or energy facility; for example, coal steam-electric, wood steam-electric,

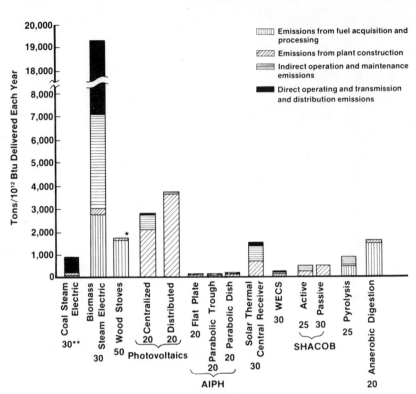

*Emissions are uncontrolled for wood stoves; all other technologies employ best available control technologies and/or meet New Source Performance Standards.

**Plant lifetime in years.

Figure 4–5. Cumulative Life-Cycle Hydrocarbon Emissions, 1985 Data

and solar-thermal plants. However, the focus of the following sections has been narrowed to address those impacts unique to the solar-energy systems modeled within SEAS. These impacts are discussed by collection method and conversion technology.

4.2.4.1 Solar-Heating Systems. Two solar heating systems are included within SEAS: passive and active. Each system is discussed separately. Passive solar heating systems require no unique materials for their construction. Only additional amounts of concrete and glass are required for construction of a building receiving heat via a passive solar design as compared to a conventional building. The emissions associated with processing concrete and glass are well accounted for within SEAS. As a consequence, there are no unique environmental effects resulting from deploying a passive

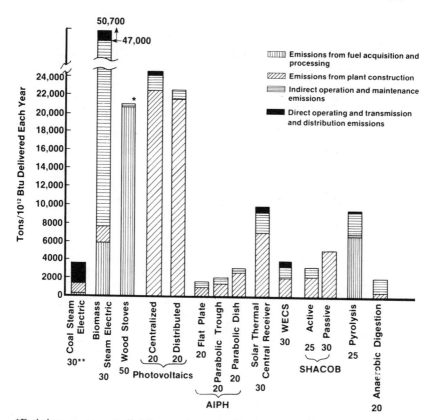

*Emissions are uncontrolled for wood stoves; all other technologies employ best available control technologies and/or meet New Source Performance Standards.

**Plant lifetime in years.

Figure 4-6. Cumulative Life-Cycle Carbon Monoxide Emissions, 1985 Data

solar heating system that are not included within the model or do not normally occur during building construction. This situation is not true for active solar building and hot-water systems. An active solar heating system includes the following subsystems: (1) flat-plate collectors; (2) heat exchangers; (3) energy storage systems; (4) pumps; (5) auxiliary heaters; and (6) pipes, controls, and valves. The working fluid for the SEAS generic system is water to which antifreeze (for example, propylene or ethylene glycol), anticorrosives (chromates, sodium nitrates and nitrites, and sulfates), and biocide compounds (usually chlorine) have been added. Most of the additives are toxic and may present environmental or health hazards if the solar heating system should overheat or burn or if working fluids leak or are dis-

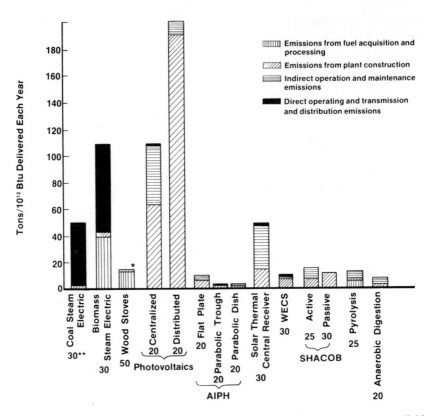

*Emissions are uncontrolled for wood stoves; all other technologies employ best available control technologies and/or meet New Source Performance Standards.

**Plant lifetime in years.

Figure 4-7. Cumulative Life-Cycle BOD Emissions, 1985 Data

charged from the system (Searcy 1978). Many of these hazards are not adequately accounted for within SEAS emissions projections. Table 4-34 presents a listing of typical additives, their toxicity, and their flammability.

Under normal operating conditions, an active solar heating system emits no air pollutants. However, if the system should overheat or burn, toxic volatile organics may be released. These organics will usually condense on the inside surface of the collector in the case of the system overheating (Energy and Environmental Analysis, Inc. 1977; U.S. Department of Energy 1978a; U.S. Energy Research and Development Administration 1977a; Searcy 1978). System burning may release hydrogen cyanide, TDO (toluene diisocyanate), hydrogen chloride, hydrogen fluoride, ammonia, nitrogen oxides, and other irritants from the combustion of plastics, other

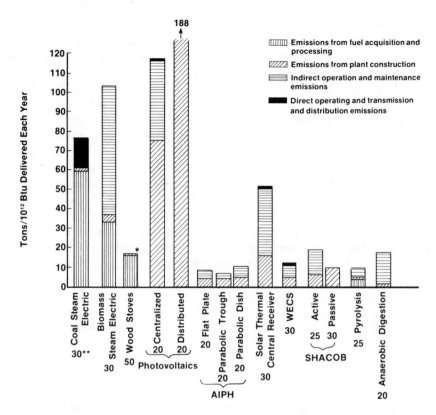

Figure 4–8. Cumulative Life-Cycle Suspended-Solids Emissions, 1985 Data

*Emissions are uncontrolled for wood stoves; all other technologies employ best available control technologies and/or meet New Source Performance Standards.
**Plant lifetime in years.

synthetics, and working fluids (Energy and Environmental Analysis, Inc. 1977; Consroe et al. 1976; U.S. Department of Energy 1978a; U.S. Energy Research and Development Administration 1977a).

Leakage, accidental release, or improper disposal of the system working fluids may present water-pollution hazards. The heat-transfer fluid potentially may enter potable water at the interface between the working fluid and heat exchanger in a hot-water system. The hazard of such a leakage depends on the concentration of toxic additives within the working fluid. Chromates and dichromates (anticorrosives for aluminum heat exchangers and pipes) are known carcinogens of the lymph system, nasal cavity, and paranasal sinuses and possibly of the stomach and larnyx. Their usual concentration in an active solar heating hot-water system is 20 to 30 ppm. Nitrate and nitrite additives, also present in small amounts in the working fluids, can decompose to carcinogenic nitrosamines. Recent studies indicate that upon leakage into a hot-water system the additives will be diluted to levels below present EPA drinking-water standards and should pose little health hazard (Energy and Environmental Analysis, Inc. 1977; Consroe et al. 1976; U.S. Department of Energy 1978a; U.S. Energy Research and Development Administration 1977a). However, ethylene glycol is present at high enough concentrations to pose a health hazard. To avoid such risks, current designs often employ double wall heat exchangers and/or add a dye to the working fluid so that leaks can be visibly detected.

The working fluid in active solar heating systems will undergo decomposition and must be periodically flushed and replaced to maintain system efficiency. Improper disposal of spent working fluids (such as dumpage into domestic storm sewers) could pose risk to local aquatic systems or municipal water systems if emission levels are high. Fluid additives could contaminate drinking water, harm aquatic species, and reduce the effectiveness of activated sludge sewage treatment facilities. Mitigation measures include installation of fluid catch basins with subsequent fluid recycling or disposal (Energy and Environmental Analysis, Inc. 1977; U.S. Department of Energy 1978a; U.S. Energy Research and Development Administration 1977a).

Other potential risks include control of vermin and vegetation through use of pesticides around collectors (depending on the site); possible structural failure of roofs, especially for retrofit systems; glare from collectors; and glass breakage due to stress from hail, snow, ice, wind, and so forth (Energy and Environmental Analysis, Inc.; U.S. Energy Research and Development Administration 1977a).

4.2.4.2 AIPH Systems. As previously discussed, three AIPH systems were modeled within SEAS: flat-plate collectors, a parabolic-trough, and a parabolic-dish total-energy system. All three systems utilize water as the working fluid. The working fluids will contain additives that are analogous to those employed in the active solar heating system.

In addition to the potential impacts discussed for active heating systems, there are risks unique to AIPH systems and their applications. Flat-plate AIPH systems typically have operating temperatures in the range of 60°C to 100°C. Parabolic troughs and dishes can achieve operating temperatures of 280°C to 400°C or greater (MITRE 1977d). Maintenance and plant personnel may suffer burns if they come in contact with collectors during operation or if working fluids are released. Release of fluid also may pose a contamination hazard if it contacts industrial or agricultural products (U.S. Energy Research Development Administration 1977b; U.S. Department of Energy 1978b).

Table 4–34
Active Solar Heating Systems: Candidate Working Fluid Characteristics
(*tons/10^{12} Btu*)

Heat Transfer Fluids and Additives	Flammability	Toxicity
Water (H)[a]	—	—
Dimethyl siloxane polymers (H)	low; flash point depends on viscosity	low
Aromatic hydrocarbons (H)	low	moderate
Ethylene glycol (H)	moderate; flash point 116°C	lethal dose humans 100 ml; TLV 10 mg/m^3 as particulate, 250 mg/m^3 as vapor
Fluorocarbons (H)	nonflammable to practically nonflammable	TLV 1000 ppm
Lithium chloride (H)	nonflammable	very toxic; TLV 0.05 mg/m^3
Lithium nitrate (A)[b]	moderate	moderate if ingested or inhaled
Lithium hydroxide (A)	nonflammable	moderate if ingested or inhaled
Sodium chromate (A)	low	highly toxic; TLV 0.5 mg/m^3

Sources: American Conference of Governmental Industrial Hygienists, 1978, *TLVS: Threshold Limit Values of Chemical Substances in Workroom Air Adopted by ACGIH for 1978,* Cincinnati, Ohio: American Conference of Governmental Industrial Hygienists; Searcy, J.Q. ed., 1978, *Hazardous Properties and Environmental Effects of Materials Used in Solar Heating and Cooling (SHAC) Technologies: Interim Handbook,* Albuquerque, N.M. Sandia Laboratories, SAND 78–0842, August; U.S. Department of Energy, 1978a, *Environmental Development Plan: Solar Heating and Cooling of Buildings, 1977,* Washington, D.C.: U.S. Department of Energy, DOE/EDP–0001, March; Windholz, M., ed., 1976, *The Merck Index: Encyclopedia of Chemicals and Drugs,* 9th ed., Rahway, N.J.: Merck and Co., Inc..
[a]H = heat transfer fluid.
[b]A = fluid additive.

4.2.4.3 Solar-Thermal-Electric System. The generic solar-thermal-electric system is a central receiver utilizing water as the working fluid and rock and oil as the storage subsystem. The technology is materials intensive (for example, see Caputo 1977, table 6–2). In addition, it is land intensive, requiring approximately 3 km^2 per 100 MW plant and support facilities (Davidson and Grether 1977; Davidson et al. n.d.). The environmental effects of processing the materials input generally are accounted for within SEAS damage projections. However, many impacts will result from the physical presence of the plant and from heliostat reflection of radiant energy.

Solar-thermal plants probably will be sited in the desert in the Southwest during the initial commercialization stage because of favorable insolation conditions. Installation of solar-thermal central receiver plants at desert sites will have impacts similar to any large construction activity. Most burrowing desert species spend daytime hours underground to escape the heat. Excavation and site grading, which will be performed during daylight hours, will probably destroy burrowing species in the immediate area. Construction activities, noise, and other activities noxious to wildlife will force emigration of mobile species and destroy many less mobile and sessile species (Davidson and Grether 1977; Davidson et al. no date; U.S. Energy Research and Development Administration 1977c; Sears et al. 1977; MITRE 1975). The weight of heavy equipment may crush some burrowing species. Tracks also will increase the potential for erosion and runoff, airborne dust, and turbidity at the site during construction due to destruction of desert crust and desert pavement (Davidson and Grether 1977).[4]

Although none of the criteria air pollutants are emitted, the plant operation will present environmental hazards. Principal hazards include potential alteration of the microclimate and heliostat glare and burn risks. The presence of large number of heliostats will modify local terrain and may alter the microclimate. Estimated ground coverage by the heliostats is 50 percent. Wind speed in the area will be decreased while surface air turbulence may increase. Shade from the heliostats is anticipated to decrease surface temperature and evaporation and thus increase relative humidity at the site. Drift from wet cooling towers also will increase the moisture level. In combination, changes in the microclimate may induce changes in the relative composition of the plant community and thereby affect the composition of the animal community. The plant will emit waste heat (5.1×10^8 Btu/hr for a 100 MWe plant operating at 100 percent capacity [Sessler and Cukor 1975]), and reflect radiant energy from the heliostats. At the same time, some thermal energy will be removed from the environment by conversion to electricity. The net thermal effects are difficult to analyze (Davidson and Grether 1977; U.S. Energy Research and Development Administration 1977c; Sessler and Cukor 1975; Sears and McCormick 1977).

Misdirected light from heliostats may present risks of fire, burn, glare, and eye damage. Present plant designs include safety precautions to minimize these risks. Precautions include exclusion zones, beam control methods, and fail safe, rapid heliostat stow mechanisms. Exclusion zones are designed to restrict both vertical and horizontal visual access to the site in case of the worst random coincidence of uncontrolled heliostat beams. The required vertical height of this zone has been estimated to be 300 meters for coincidence of two heliostat beams, and 900 meters for fifty beams (U.S. Department of Energy 1978c). Beam control methods are designed to limit coincidence of a specified number of heliostat beams beyond a certain altitude. Thus aircraft flying above this altitude would never be exposed to more than one heliostat beam at any given instant. Fail-safe stow mechanisms have been developed to protect against misdirected heliostat beams in the event of malfunction of the computerized heliostat tracking subsystem. Random beam movement could present eye damage and fire risks. Hence the fail-safe system would rotate the heliostats to their stow position within five seconds of computer malfunction (Brumleve 1977; U.S. Department of Energy 1978c; Boeing Engineering and Construction Co. 1977).

4.2.4.4 Photovoltaic Systems. Two photovoltaic-system designs were modeled within SEAS: a centralized system and a dispersed system. Both employ copper-sulfate–cadmium-sulfide cells (Cu_2S/CdS). Environmental risks occur throughout the life cycle. Of particular concern are (1) toxic emissions resulting from cell production; (2) release of toxic fumes should cells burn; (3) dangers associated with ingestion of cell fragments, for example, by a child; and (4) precautions required upon system decommission. (Effects of siting centralized photovoltaic systems in desert environments are very similar to those associated with solar-thermal-electric plants.)

Several methods are utilized currently for production of Cu_2S/CdS. In one technique, zinc is electroplated over a layer of copper foil. Next, CdS is vacuum deposited over the zinc. The Cu_2S layer is produced via ion exchange using copper chloride as the starting material. The cells are then hermetically sealed in tempered glass to prevent cell oxidation (U.S. Energy Research and Development Administration 1977b; Fann 1978). In addition to the 0.2-pound cadmium particulates emitted per ton cadmium mined (U.S. Environmental Protection Agency 1977), cadmium and cadmium compounds will be emitted from the manufacturing process. Manufacture of 10,000 MW peak cells per year may result in the release of thirty-four tons of cadmium during cadmium-metal production; fifty-two tons of CdS from CdS production; and unspecified quantities of cadmium, cadmium oxide, and toxic electroplating compounds from cell production, depending on cell and production design. It should be noted, however, that cadmium releases from manufacture of 10,000 MWe cells is about 5 percent of the cadmium emissions resulting from zinc refining alone (U.S. Energy Research and Development Administration 1977b; U.S. Department of Energy 1978e). In addition, production is largely a laboratory process at present. Thus estimates of processing emissions are speculative.

Because cadmium is toxic, emissions are tightly controlled under Environmental Protection Agency (EPA) guidelines. Cadmium is bio-accumulated from air and water, but there is no evidence of bioconcentration in aquatic food chains. Cadmium is both an acute and chronic toxicant to humans. Inhalation of large amounts of cadmium oxide (CdO) can produce pulmonary edema and death. Additional acute effects of CdO include throat dryness, vomiting, cough, chest pain, and pneumonia may develop in some cases. Chronic exposure to low levels of cadmium induces renal tubule damage, proteinuria, abnormal renal vasculature, and emphysema. Cadmium has also been implicated as teratogenic and as a factor in hypertension (U.S. Energy Research and Development Administration 1977b; Fleischer 1974; Olsen et al. 1973).

Cu_2S/CdS systems ordinarily do not operate at elevated temperatures and consequently do not require a cooling system. However, should cells burn, toxic fumes will be released from the combustion of the cells and synthetics used in current designs (U.S. Energy Research and Development Administration 1977b). Estimations of toxic-fume releases have been performed for what the authors defined as the worst-case situation: Cu_2S/CdS cells on the roof of a wood frame, single family, and residence. A total of 2,421 ft² (225 m²) CdS cells (CdS layer 30μ thick; corresponding to 62 lb [28 kg] CdS per house) was assumed for the residence. It was further assumed that enough oxygen was available to convert all the CdS to CdO. Upper-bound estimates for CdO were 0.15 ppm (184 g/m³) (Olsen et al. 1973) or approximately four times the TLV[5] of 0.05 mg/m³ (American Conference of Governmental Industrial Hygienists 1978), 11 to 16 yards (10 to 15 meters) from the fire. Carbon-monoxide concentrations at this distance were estimated to be 2500 ppm, well above the current TLV of 50 ppm. CdO concentration 27 yards (25 m) downwind was calculated to be 1.6×10^{-3} ppm (1.96 g/m³). The authors estimated that 90 percent of the CdS in the cells would end up in the ash in a form combined with other metals (Olsen et al. 1973).

In the same study, Olsen et al (1973) estimated the risk associated with ingestion of the Cu_2S/CdS cells. Cadmium is an emetic and, therefore, Cu_2S/CdS cells or cell fragments would probably be regurgitated if ingested. If not, cell ingestion could cause illness or prove fatal if ingested in large enough quantities. The authors estimated that a child would have to eat .75 in² (5 cm²) of cells before a severe reaction would be elicited. Risk from cell ingestion was judged thus to be of low probability.

The final point of potential adverse environmental effects occurs on decommission of the Cu_2S/CdS system. Because of their construction (that is, a thin layer of Cu_2S over a 5- to 30-micron thick layer of Cds, Cu_2S/Cds cells cannot be recycled, given current technology (U.S. Energy Research and Development Administration 1977b). Incineration of cells is not an acceptable disposal option because of the release of toxic cadmium compounds upon cell combustion. Cells can be landfilled, but only at sites not subject to acid drainage since leaching of cadmium into waterways can produce environmental poisoning (U.S. Energy Research and Development Administration 1977b; Fleischer 1974).

4.2.4.5 Wind Energy Conversion Systems. As discussed in section 3.1, WECS examined with SEAS is a 1.5 MWe unit deployed in a centralized mode; that is, in *wind farms*. Materials required for unit fabrication are not exotic.

Although WECS emit no air, water, and thermal pollutants or solid wastes during the operation phase, large wind farms may produce environmental impacts. The local microclimate may be altered due to the physical presence of the WECS and from changes in local wind-flow patterns. The WECS structure is expected to produce environmental effects similar to those attributable to windbreaks; that is, decrease local wind speed and increase turbulence. Theoretical calculations indicate that the speed of moderate winds (9 to 14 mph [4 to 6 m/sec]) will be reduced by a maximum of 25 percent. Changes in wind-flow patterns will affect relative humidity, vapor pressure, temperature, and evaporation rate. Changes in microclimate, if persistent, in turn will induce changes in the relative composition of plant and animal communities (U.S. Energy Research and Development Administration 1977d; Rogers et al. 1976). The degree of WECS impact on microclimate currently is not known and field investigations are underway (U.S. Department of Energy 1978d).

The presence of WECS may also present an obstacle hazard to flying species. Songbird species typically migrate at altitudes between 420 to 1480 ft (150 to 450 m) but adjust height according to weather and terrain (for example, ridges are crossed at altitudes below 420 ft). Thus hazards to some bird species may be minimized through proper site selection, such as not atop ridges (Brumleve 1977). The collision probability has been estimated for birds entering the rotor-disk area of a WECS unit assuming no evasive action is taken. The collision probability for a bird with flight speed of 8 mph (3.5 m/sec) is 13 percent and for flight speed of 18 mph (8 m/sec), 8 percent. Collision probability decreases as flight speed increases. Average songbird flight is 16 to 27 mph (7 to 12 m/sec) and that of water fowl and shorebirds is 34 to 56 mph (15 to 25 m/sec) (Rogers et al. 1976). Birds flying through the disk area of the MOD-O WECS unit have been observed to take evasive action (U.S. Department of Energy 1978d).

WECS may also present hazards to workers and public safety through (1) structural failure of the tower; (2) blade or blade-tip throw; and (3) obstruction of air space. Tower and blade failure can result from mechanical stress, rotational forces, wind sheer, or cataclysmic weather events. The hazard zone for tower failure is a circle with a radius equal to the tower height. The hazard area for blades is much larger and is estimated to be approximately one-fourth mile for a 1.5-MW horizontal axis WECS (U.S. Energy Research and Development Administration 1977d; U.S. Department of Energy 1978d). All structures over 200 ft (61 m) must be marked according to Federal Aviation Administration tall-structure regulations; thus, the WECS would be marked. A tall structure is not considered to be an air hazard until it is 500 feet or more high (Black & Veatch Consulting Engineers 1978).

4.2.4.6 Silvicultural Farms. Silvicultural farms will produce environmental effects similar to other large silvicultural and agricultural operations. Effects include (1) impact of emissions from cultivation and harvest equipment on local air quality; (2) runoff, dust, and so forth from extensive site preparation to remove competitive species; (3) release of fertilizers and pesticides often required to maintain growth of monoculture plantings; (4) potential depletion of soil nutrients; and (5) possible degradation of local water quality. The most important of these, soil nutrient depletion and water quality degradation, are discussed here.

Nutrient depletion largely depends on whether residues are left on the farm and the number of cuts (or harvests) made per rotation. Removal of the whole tree will deplete soil nutrients at the site more than removal of the main body of the tree only; that is, leaving some residues at the site (U.S. Energy Research Development Administration 1977f; Hall et al. 1976; Dunwoody 1978). In addition, repeated large cuts tend to cause a greater annual nutrient drain than a single cut per rotation. Residue decay, however, is only one pathway by which nutrients enter the soil. Other pathways include mineralization of the solum and the secondary materials in the soil and atmospheric inputs in the form of particulates and dissolved materials in precipitation. Thus it may be possible to supplement soil nutrients with fertilizers or nitrogen-fixing plants. However, fertilization is not believed to be a management tool to offset any long-term effect of nutrient removal due to harvesting practices (Hall et al. 1976; Povich 1977).

Water quality can be degraded via thermal pollution and runoff of sediments and discharges associated with agricultural practices (fertilizers and pesticides). The amount of runoff will be influenced by the degree of soil disturbance at the site. If the soil is left basically undisturbed, it will retain a high water-absorption capacity. However, destruction of surface vegetation cover and severe soil disturbance (for example, clearing of logging roads) will not only increase the propensity for wind erosion but also water erosion and removal of surface soil layers, and it will induce sedimentation loading and turbidity of local streams (U.S. Energy Research and Development Administration 1977f; Hall et al. 1976; Bennett et al. 1973).

Canopy removal will also increase runoff and erosion. The canopy typically decreases wind speed and the velocity at which precipitation strikes the soil surface. As a result of canopy removal, nearby stream flow may be increased. Removal of the canopy could also increase the temperature of streams flowing through the farm. Stream temperature increases up to 8.3°C have been recorded in some clearcut areas. The amount of temperature increase is a function of the rate of stream flow and the length of stream exposed to the sun. Temperature increases can be mitigated or avoided by leaving a buffer strip of trees near the stream (Hall et al. 1976; Bennett et al. 1973).

4.2.4.7 Agricultural and Forest Residue Collection. Residues perform a variety of environmental functions: (1) help to control wind and water erosion; (2) maintain soil tilth; (3) maintain fertility of soil; and (4) preserve soil moisture. Removal of residues can have both positive and negative effects.

Utilization of biomass residues does not require the commitment of land to the production of fuel crops. Residues generated through agricultural and silvicultural practices, therefore, become a fuel resource. Thus a biomass-fuel crop can be obtained without competing with food, fiber, and forestry demands for land (U.S. Energy Research and Development Administration 1977f; Dunwoody 1978).

Removal of excess amounts of residues can benefit both fire and pest protection activities. The frequency of fires is seven times higher and more severe in forests in which residues remain than in forests not containing large amounts of residues. Residues may often harbor forest and agriculture pests. Epidemics of bark beetles have been recorded as starting in forest residues (Howlett and Gamache 1977). Agricultural residues often provide overwintering shelter to pest species such as the boll weevil. Thus removal of residues may have a positive impact on pest control operations.

Recent investigations indicate that for some crops, leaving too much residue on the field may have a negative impact on soil nutrients. For example, soil species that decompose wheat straw require and consume much of the deposited nitrogen fertilizer to break down the straw. The straw, however, has very little soil nutrient value. Therefore, fewer total nutrients are returned to the soil than are consumed to degrade the straw.[6]

If forest residues are excessive, they often clog streams draining the area. Residue clogs not only obstruct stream flow but also deplete oxygen by increasing BOD during degradation. Affected streams may be less desirable wildlife habitats. Large amounts of forest residues may restrict the movement of game animals, depress the growth of forage crops, and decrease the recreational attractiveness of the area. Removal, therefore, may have a positive effect for large wildlife species and improve the aesthetic attractiveness of the forest site (U.S. Energy Research and Development Administration 1977f; Bennett et al. 1973; Howlett and Gamache 1977).

Removal of residues can also have negative impacts, particularly if all residues are collected. Collection of too much of the residues from an agricultural field increases airborne dust emissions and water-induced runoff. Airborne erosion may be reduced somewhat by leaving some of the residues behind or by use of low-till agriculture (U.S. Energy Research and Development Administration 1977f; Bennett et al. 1973; Howlett and Gamache 1977). Erosion not only increases atmospheric and water turbidity and stream sedimentation, but it may serve to carry insecticides, herbicides, and fertilizers from the site (U.S. Energy Research and Development Administration 1977f; Howlett and Gamache 1977; Davidson et al. 1977). Residues reduce water erosion primarily through interception of raindrop impact and reduction of runoff water velocity. Residues reduce wind erosion by decreasing wind speed and preventing direct wind forces from reaching erodible soil particles (Lindstrom et al. 1978). Soil erosion by water may be severe if surface cover on steep forest slopes is disturbed as would occur if stumps and roots are also removed during forest-residue collection (Howlett and Gamache 1977). Thus only limited amounts of residues can be removed without producing severe erosion.

Another area of concern associated with residue collection is the potential for soil nutrient depletion and damage to soil quality. As previously mentioned, decomposition of residues is one of the three pathways by which nutrients enter the soil. Complete removal may deplete soil nutrients and necessitate the use of fertilizers, especially if foliage is also collected. Heavy planting and harvest equipment could cause severe soil compaction thus impeding aeration, water infiltration, and productivity for many years (U.S. Energy Research and Development Administration 1977f; Bennett et al. 1973; Howlett and Gamache 1977).

4.2.4.8 Anaerobic Digestion. The anaerobic digestion system modeled with SEAS (described in section 3.1) utilizes cattle manure as the starting feedstock. No exotic materials are required for system fabrication, as illustrated in sections 4.2.2 and 4.2.3. However, there are potential environmental problems unique to anaerobic digestion. Included are manure collection and preparation, waste gaseous emissions from the plant, disposal of the solid waste products, and the treatability of effluents from gas clean-up operations.

Manure is collected from feedlots for sanitary reasons regardless of its subsequent use. Presently, manure is used as a fertilizer. Water requirements for collection of manure for use in digestors are not expected to differ significantly from amounts used for cleaning purposes.

The anaerobic digestor itself will emit some waste gases. Most are accounted for by the SEAS model. Emissions include carbon dioxide, sulfur dioxide, and some nitrogen dioxide. The sulfur and nitrogen gases are derived from the approximately 1 percent hydrogen sulfide and ammonia in the raw biogas (Ballou et al. 1978; U.S. Energy Research and Development Administration 1977f).

A major environmental question is the disposal of waste products from the digestor; that is, effluents from the digestor itself, gas clean-up equipment, and solid wastes. Disposal options for the liquid effluents include (1) discharge to municipal water-treatment systems that may stress treatment capabilities; (2) landfill of the effluent along with the solid wastes; and (3) use of the effluent for irrigation immediately after discharge or use after containment in an evaporation pond containing algae (either irrigation option may require dilution of the effluent because of high salt content). Disposal options for the sludge, or solid wastes, include: (1) landfill; (2) use of the sludge as a soil fertilizer and conditioner; and (3) recycling of the sludge as a cattle refeed. Use of the sludge as a fertilizer has several advantages. First, the original use for manure, fertilizer, is met thus alleviating the need for synthetic, replacement fertilizers. Second, sludge used as a fertilizer does not offer as great a runoff hazard as raw manure. However, runoff hazards will exist especially when applying the sludge during winter months (Roop 1978). The sludge and liquid effluents are fairly high in protein, which may make them good candidates for refeed (Hashimoto et al. 1978). The potential concern associated with using the sludge as fertilizer or refeed is due to the concentration of heavy metals. Metals typically found in manures are shown in table 4-35. Further research is needed to determine the environmental fate and effects of the heavy metals when digestor sludge is used as fertilizer or refeed (Ballou et al. 1978; U.S. Energy Research and Development Administration 1977f; Hashimoto et al. 1978).

4.2.4.9 Pyrolysis of Residues. A Purox® pyrolysis system producing medium-Btu gas was modeled using SEAS. The feedstocks (forest or agriculture residues) are heated at a very high temperature in the presence of

Table 4-35
Heavy Metal Content in Anaerobic Digestor Sludge
(*tons/10¹² Btu*)

Metal	10^3 Tons/10^{12} Btu Energy Produced
Iron	0.609
Aluminum	0.568
Manganese	0.073
Zinc	0.071
Lead	0.010
Copper	0.0095
Cadmium	0.0007

Source: Adapted from Steve Ballou et al., "Environmental Residue and Capital Cost Evaluation for Energy Recovery from Municipal Sludge and Feedlot Manure," Draft Report, Argonne, Illinois: Argonne National Laboratory, August 1978.

oxygen. All or almost all of the organic matter in the feedstock is either oxidized or pyrolyzed to a medium-Btu gas. No organics or chars remain, and production of oil is minimized or nearly eliminated.

Three important waste streams result from the Purox® process: (1) wastewater condensed from the converter offgas; (2) slag or solid waste; and (3) cooling tower blowdown. The gas from the pyrolysis process will contain relatively few sulfur and nitrogen gases because of the low initial content in the feedstock. Sulfur content is almost always less than 0.3 percent per dry unit weight and ash content is generally less than 1 percent per dry unit.[7] The water condensed from the gas will contain appreciable amounts of light organics: C_1 to C_4 aliphatic compounds and low molecular weight aromatics such as phenols, benzene, and furans. At present recovery is impractical. Some of the metals present in the residue feedstocks will be volatized during pyrolysis and probably end up in the wastewater. The wastewater may also contain significant amounts of chlorine and zinc. Reducing chlorine and zinc to levels below drinking water standards may be a problem depending on the feedstock (U.S. Energy Research and Development Administration 1977f; Gikis et al. 1977).

The amount of slag waste remaining after pyrolysis is dependent on the feedstock. Estimates of solid wastes generated are 1.6 lb/10^6 Btu for wood wastes as the feedstock, 10.6 lb/10^6 Btu for barley and cotton wastes, and 34.7 lb/10^6 Btu for rice wastes. The slag could be landfilled and impounded to prevent leaching, used for industrial purposes (for example, as rubber reinforcement or as a cement additive), or used as a fertilizer because of its potash and phosphate content (Gikis et al. 1977).

4.2.4.10 Steam-Electric Wood Combustion. Air emissions resulting from the combustion of wood in large boilers to produce electricity include particulates (controllable through techniques used for coal combustion), sulfur compounds (at levels below coal), nitrogen oxides (more than from coal combustion), carbon monoxide, and hydrocarbons (U.S. Energy Research and Development Administration 1977f; Hall et al. 1976). These air emissions are accounted for within the SEAS model.

In addition, other organic compounds are emitted from the wood boiler. These emissions are mostly low molecular weight hydrocarbons and alcohols, acetone, simple aromatic compounds, and several short-chain unsaturated compounds such as olefins. Emission of carcinogenic compounds is suspected but is at present unknown. Photochemically active compounds may be emitted under poor combustion conditions. The inordinately high amounts of hydrocarbons released under these conditions can react in the presence of sunlight to form peroxy compounds. The peroxy compounds can interfere with the normal nitrogen-oxygen cycle by oxidizing nitric oxide to nitrous dioxide, thereby preventing the destruction of ozone. Unsaturated hydrocarbons are the most photochemically reactive of the compounds released. Olefins with less than four carbons, like those formed during combustion of wood, are much less reactive. However, emissions of photochemically active hydrocarbons are low even under poor combustion conditions (U.S. Energy Research and Development Administration 1977f; Hall et al. 1976).

The principal solid waste generated from combustion of wood is ash. Assuming a precombustion ash content in the wood of 1.5 percent, an estimated 8 lb of ash will be generated per 10^6 Btu delivered energy.[8] The ash will contain trace metals derived from the trace metals in the wood feedstock. Trace compounds typically found in the ash are listed in table 4-36. Disposal options for the ash include landfill or use as a soil conditioner and possibly a nutrient supplement (U.S. Energy Research and Development Administration 1977f; Hall et al. 1976). The experience of the paper industry in controlling emissions and wastes from wood combustion should provide guidance in developing pollution control options.

Table 4-36
Trace Metal Content of Wood-Derived Ash
(*tons/10^{12} Btu*)

Compound	Concentration, ppm
Silicon	19.6
Aluminum	3.6
Calcium	2.9
Sodium	2.1
Magnesium	0.8
Potassium	0.3
Titanium	0.1
Manganese	0.016
Zirconium	0.006
Lead	0.003
Barium	0.010
Strontium	0.002
Boron	0.003
Chromium	0.001
Vanadium	0.001
Copper	0.001
Nickel	0.001
Mercury	nil

Source: E.H. Hall et al., *Comparison of Fossil and Wood Fuels,* Washington, D.C.: U.S. Environmental Protection Agency, March 1976, EPA-600 2-76-056, p. 80.

4.2.4.11 Summary. Most of the solar systems modeled using SEAS do not emit air or water pollutants during their operation. However, their deployment will have environmental effects. Working fluids may be released (via accidents or improper disposal techniques) from active solar heating, AIPH, and solar-thermal systems. This release could pose risks to humans and wildlife. Operation of WECS may pose a slight hazard to flying species and produce low-level noise pollution. Installation of centralized photovoltaic and solar-thermal central-receiver facilities at desert sites will affect the local ecosystem and may alter the immediate microclimate. The biomass technologies will emit air pollutants but, in general, their effects are captured within the SEAS model. Collection and production of biomass feedstocks may deplete soil nutrient levels in addition to creating the erosion problems estimated within SEAS. Operation of biomass conversion facilities will produce wastes (ash, char, and so forth) but at quantitatively lower levels than from coal combustion facilities. However, because of the absence of or low emissions in the operating phase, the solar-energy facilities will have lower life-cycle effects than conventional energy conversion options.

4.3 Policy Implications

The principal general conclusion reached by this book is that solar-energy technologies can be used to reduce the environmental insults generated by the energy sector of our economy. During the 1975–2000 period, however, this potential reduction in environmental insult is not nearly as large as that directly obtainable through the application of environmental control technology. Nor is the environmental benefit potentially obtainable through the deployment of solar-energy automatic.

In general, solar technologies have large environmental impacts associated with their initial construction, due to the large amount of material used per Btu of energy delivered, relative to conventional systems. On the other hand, the solar technologies have minimal impacts associated with the operating phase. This clearly implies that the construction of a solar-energy facility represents an environmental investment that is not recovered until some years of operation have passed. Concomitantly, during a period of rapid investment in solar facilities, the environmental benefits of solar energy, per Btu, will be lower than in the steady state.

The time path of environmental benefits from solar energy has profound policy implications due to the inherent intergenerational conflict involved. By maximizing the rate of solar deployment in the present period, we hasten the time when the percentage of energy supplied by solar systems has reached a steady state; thus the time is hastened when society begins to reap fully the potential environmental benefits available through the deployment of solar energy. On the other hand, during a period of rapid growth in the deployment of solar-energy systems, the environmental benefits available are temporarily lower than what they would be under slower deployment conditions. Under the Domestic Policy Council's high solar scenarios the deployment of solar technologies grows at 35 percent per year from 1975 to the year 2000. This clearly implies that the absolute investment in solar-related construction is largest during the year 1999–2000 and environmental insults related to this construction are also at a maximum. At some point, presumably soon after 2000, the rate of growth of deployment of solar technologies must taper off as the steady-state situation is approached and the potential environmental benefits of solar energy systems will be fully realized.

If the Domestic Policy Council's solar deployment goals for the year 2000 are to be met, and the generation living as the year 2000 is approached is not to be environmentally shortchanged at the expense of post-2000 generations, the policy must favor more rapid growth in the deployment of solar-energy technologies in the early part of the 1975–2000 period with a tapering off in the rate of growth toward the latter part of that period. The alternative is reduced growth in the deployment of solar technologies in the few years prior to 2000, with larger deployment thereafter, leading to a deferral of the DPR solar goals. Figures 4–9 and 4–10 illustrate these issues. Figure 4–9 shows the DPR target solar deployment for the year 2000 under the maximum-feasible case, assuming a constant 35 percent per year rate of growth. Figure 4–10 shows the associated absolute growth in solar deploy-

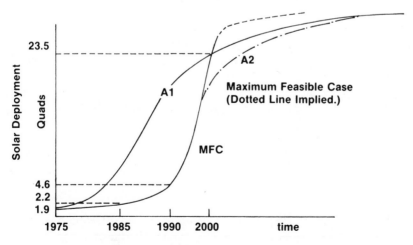

Figure 4–9. Solar-Energy Deployment as a Function of Time

ment during the same period. In these diagrams, the maximum-feasible case is labeled MFC. The early deployment policy alternative, which preserves the DPR solar goals for the year 2000, is labeled A_1. The alternative, which reduces the DPR solar goals, is labeled A_2. A large component of solar-related environmental insult is correlated with the absolute growth curve, shown in figure 4-10, since construction impacts are related to absolute increases in solar deployment. The early deployment growth path (A_1) has clearly different intergenerational environmental implications than the DPR deployment path. (As an aside, the early solar deployment path has one important nonenvironmental advantage over the DPR scenario shown. Since the variations in the amount of solar-energy facilities constructed each year are much smaller in the alternative scenario, industry will have a much easier time gearing up and down to produce the facilities.)

A number of other policy implications emerge. An important conclusion to be drawn is that solar-energy technologies could be useful tools for environmental control in selected regions. If the solar technologies whose operating phases are most environmentally benign (solar heating and cooling of buildings and photovoltaics, for example) were deployed in areas where the current technologies are based on oil or coal combustion and, at the same time, if the construction of the solar-energy facilities could be located outside the region, an environmental transfer payment from one region to another would be effected. This conceivably could be done as a matter of deliberate policy in constricted air sheds such as the South Coast Air Basin in southern California.

Section 4.2 indicates that not all solar technologies are environmentally benign even in the operating phase. Wood stoves are a major contributor to particulate emissions under the DPR scenarios. In some regions (like the Northeast) wood stoves dominate particulate emissions to the atmosphere in the year 2000. This has dramatic implications for public policy. Either a cost-effective technology to control particulate emissions must be found or the use of wood stoves must be limited in regions with air-quality problems or meteorological conditions that would be averse to large particulate emissions. The latter policy recently has been adopted in Vail, Colorado, where it was found that during the winter ski season, which is subject to frequent air inversions, particulate emissions from wood stoves are a real problem.

Silvicultural farming is the other solar technology with major environmental effects associated with its operating phase. Here the problem is agricultural runoff. The solution will require careful monitoring of runoff combined with further research into soil loss prevention practices such as no-till farming.

The major indirect environmental effect of the solar technologies taken as a group is the emission of particulates from the stone-and-clay industry, which is stimulated by the construction requirements for a number of the

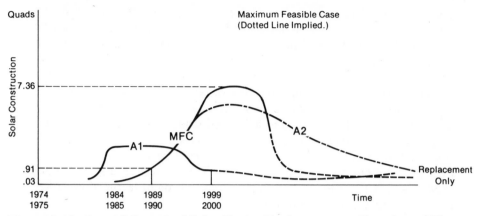

Figure 4–10. Rate of Growth of Solar-Energy Deployment as a Function of Time

solar technologies. Since the stone-and-clay-products industry is already regulated under the Environmental Protection Agency's New Source Performance Standards, both research and policy, therefore, must strive to reduce the quantities of stone and clay products used in the construction of solar facilities.

This conclusion is actually a general one: to the extent possible, research and policy should strive to reduce the materials intensity of the solar technologies. While this is, in fact, the object of much engineering research in the solar field, the environmental effects of materials-intensive production technologies imply that materials-reducing research and policy should be carried beyond the point which would be optimal from an engineering-economics perspective especially for materials whose production produces significant emissions.

4.4 Further Research

Further research into the environmental benefits and costs of solar energy can be divided into three areas: (1) approaches different than the ones used in this study; (2) extensions of this analysis; and (3) technical improvements to this study. These improvements are discussed below. Methodological improvements to the benefit-cost analysis are discussed in more detail in a previous report (Yokell 1978).

Despite the fact that this national study was highly regionalized, it was not site specific. Since actual energy-related environmental damages are often highly site specific, it would be fruitful to compare the relative impacts of solar and conventional sources of energy at particular sites. If the sites chosen are ecologically representative, some extrapolation of the results from a relatively small number of sites to the national level may be possible. A detailed economic assessment of the damages associated with the site-specific environmental impacts could then be carried out. Site-specific analysis is the most important alternative methodology, which should be used to supplement and verify the results presented in this book.

Several important extensions to this analysis could be made. First, much effort should be devoted to a detailed study of the health, property, crop, recreational, aesthetic, and other damages associated with the environmental stresses projected in this study. These impacts are implicit in the damage functions for each pollutant described in section 2.3, but more explicit and better use of the available physical damage function (dose-response) literature needs to be made.

The second major extension to this analysis needed is a procedure to calculate the pollutant emissions associated with a high, but not expanding, level of solar-energy use. Two procedures are possible. The most realistic, but difficult, approach would be to extend the SEAS model until perhaps year 2025, when the use of solar energy likely will be substantially greater than it will be in the year 2000, though growing less rapidly. The second procedure, less realistic but also less difficult, would be to run the model with an artificially elevated level of solar-energy use in the year 1985, comparable to the year 2000 solar-energy use, and a reduced growth rate between 1985 and 2000, compared with the maximum-practical scenario. This scenario would overstate the likely environmental impacts of solar energy in the early years but the year 2000 impacts using the artificially elevated level might give a more realistic picture of the steady-state impacts after the year 2000.

A number of technical improvements could be made to this work. Possible improvements to the emissions modeling effort are: (1) more or different scenarios; (2) more attention to the regional distribution of both conventional and solar-energy sources; (3) explicit incorporation of operation and maintenance of energy facilities in SEAS; (4) better input data; and (5) more input data.

As noted in section 2.1.4.2, six additional energy scenarios were developed for this analysis but not run. These scenarios incorporate different assumptions about the future price of conventional sources of energy and lead to different levels of total-energy demand, solar-energy use, and a different mix of solar-energy technologies. Each of these scenarios would provide valuable additional information to policy makers.

The regional allocation of conventional and solar-energy sources used in all the scenarios examined could be developed in a more realistic and sophisticated way than used here and described in section 2.1.4.5. This analysis used primarily a top-down approach to regional disaggregation of national scenarios. A better approach would be to create scenarios at the regional level, aggregate them to the national level, assess the national feasibility of the scenario, and then revise the regional allocations accordingly. This iterative national-regional scenario development process might provide more realistic scenarios.

In the SEAS model, no O&M expenditures are accounted for unless they are used by an INFORUM sector. Since separate input-output sectors are not built for each energy technology, some energy-related operating and maintenance expenditures are omitted. A simple program could be developed that would allocate these O&M expenditures to final demand. This would allow the incorporation of the indirect environmental effects of O&M expenditures into the analysis of energy scenarios.

Better data of two types could be used in the SEAS solar runs. First, improved capital-cost data for the energy investment module could be used for some solar technologies, particularly biomass technologies. Second, the direct emissions data used in RESGEN could be improved, again especially in the biomass area.

Finally, additional energy systems, options, and configurations could be included in the analysis. Ocean thermal-energy conversion systems, solar-power satellite systems, and liquid fuels from biomass could be analyzed.

In the economic analysis of the emissions consequences of the solar technologies, much remains to be done. First, substantial improvements to the simple rollback model of pollution dispersion could be made. Second, a more exhaustive study of the literature on environmental dose-response relationships would allow a revision and improvement of the damage functions that could be used in the BENEFITS model. Third, values-at-risk used in BENEFITS could be disaggregated into persons, property, crops, buildings, and so forth. At the present time, people serve as a proxy for all values at risk. Finally, aggregate national environmental damages used in the BENEFITS model could be updated by inclusion of currently omitted health effects such as environmentally related cancers.

By using alternative methodologies and extending and improving this analysis, improved information on the environmental benefits and costs of solar energy could be obtained. Information on other types of social benefits and costs of solar energy—occupational safety, national security, macroeconomic, labor force, lifestyle, and similar benefits and costs—may be equally critical to the policy decisions on solar and conventional sources of energy that are, and will continue to be, taken during the coming decade.

In spite of the significant improvements that could be made to this study, the author does not feel that the conclusions developed here would be substantially altered by methodological improvements. Improved pollution-damage functions would provide greatly enhanced confidence in the benefit-cost analysis but would be unlikely to change the general conclusion that the environmental benefits of solar-energy deployment are substantial. The estimated distribution of benefits and costs by region, pollutant, and industrial sector could, of course, change.

Notes

1. Because the model does not extend beyond the year 2000, there is no way to calculate the effect of truncating benefits in that year. Whether the effect is important depends primarily on the assumed social rate of discount. If a 10 percent social rate of discount is assumed, then a benefit of $1 which occurs twenty-five years hence is worth roughly $.10 in the present. If a 2.5 percent social rate of discount is assumed, this benefit is worth roughly $.50 in the present.

2. In a recent and comprehensive literature review, Fisher et al. (1979, vol. II, part II) settled on $303,000 (1977 dollars) as a reasonably conservative estimate of the statistical value of a human life.

3. The emissions released from fuel acquisition and processing and indirect operation and maintenance activities are accounted for within the scenario analyses, though they are not displayed separately from the emissions of the economic sectors which generate them.

4. Desert crust is a conglomerate of fine particles up to 6 mm thick that protects against erosion, especially by wind. It is delicate and fragile and penetrable by water. Desert pavement is a surface of fairly densely packed pebbles and stones formed after wind and water carry away finer particles of sand, silt, or clay (Davidson and Grether 1977).

5. TLV is the threshold limit value: the concentration for a normal eight-hour workday or forty-hour workweek to which nearly all workers may be repeatedly exposed, day after day, without adverse effects (American Conference of Governmental Industrial Hygienists 1978).

6. Dr. Thomas Reed, SERI Bio/Chemical Conversion Branch, 1977: personal conversation.

7. For comparison purposes, coal generally contains 10 percent ash and 0.9 to 3.0 percent sulfur (U.S. Energy Research and Development Administration 1977f).

8. By contrast, coal combustion will generate 11 lb ash/10^6 Btu and oil combustion 0.1 lb ash/10^6 Btu (U.S. Energy Research and Development Administration 1977f).

Appendix A:
Quality of the Data Sources

In any large study the quality of the data sources used varies enormously. This study generated two basic data sets in addition to the data already embedded in the SEAS model and AMBIENT/BENEFITS models. The data sets were capital-cost and operating-and-maintenance-cost vectors and residuals for the solar technologies. The data is ranked in quality in tables A-1 and A-2, so that the reader can judge the validity of conclusions based on the solar data.

Table A-1
Quality of the Solar-Residuals Coefficients Data

	Residuals to			
	Air	Water	Land	Labor [a]
SHACOB				
Passive	— [b]	—	—	2
Active	—	4	4	3
AIPH				
Flat plate	—	5	5	4
Parabolic trough	—	5	5	4
Parabolic dish	—	5	5	4
Solar thermal	—	1	2	—
Photovoltaics				
Distributed	—	1	—	5
Central	—	1	2	5
Wind	—	—	2	5
Biomass collection				
Animal residues	5	5	5	5
Agricultural/forest residues	2	—	1	4
Silvicultural farm	3	3	3	4
Biomass conversion				
Anaerobic digestion of manure	3	3	3	3
Pyrolysis of agricultural residues	1	—	1	4
Steam electric	3	2	1	3
Wood stoves	4	—	3	5

1 Based on empirical research with well-defined system parameters; highest quality.

2 Based on empirical research but some parameters too variable to define accurately (for example, transportation distances vary between systems or environmental impacts are site specific).

3 Based on paper engineering studies reviewed and updated by subsequent workers.

4 Based on educated guesses by experts in the field or on existing but very sketchy data.

5 Strictly guesswork; lowest quality.

[a] Includes operational labor, injuries, and workdays lost.

[b] Line indicates no residuals projected.

125

Table A–2
Quality of the Energy Investment Module Data for 1985

Technology	Capital Cost	Capital Cost Vector	O & M Cost	O & M Cost Vector
SHACOB—Passive	1	1	2	2
SHACOB—Active	2	3	2	3
AIPH—Flat plate	2	3	3	4
AIPH—Parabolic trough	2	2	3	4
AIPH—Parabolic dish	2	2	3	4
Solar thermal	2	2	3	4
Photovoltaics—distributed	3	3	4	5
Photovoltaics—centralized	3	3	4	5
Wind	2	3	4	5
Biomass collection—manure	4	4	5	5
Biomass collection—residues	3	5	3	4
Biomass collection—farms	3	4	4	4
Anaerobic digestion	2	3	2	3
Pyrolysis	3	4	3	4
Steam electric	2	2	2	3
Wood stoves	1	1	1	2

1 Highest quality; system is purchasable today.
2 System is purchasable today, but design changes and cost reductions have been projected.
3 Derived from a well-documented study.
4 Involves substantial guesswork on the part of the original study team or on SERI's part.
5 Assumptions based on analogy; lowest quality.

Appendix B:
Calculations Used in the
AMBIENT and
BENEFITS Models

The following equation (a modified rollback equation) is used to calculate ambient air concentrations. A prime ($'$) indicates forecast values.

$$C'_{ij} = \left\{ \frac{\sum_k P_{ijk} E'_{ijk}}{\sum_k P_{ijk} E_{ijk}} (C_{ij} - b_j) \right\} + b_j \qquad (B.1)$$

where C_{ij} (C'_{ij}) are base year (forecast) ambient concentration levels,

P_{ijk} are source-receptor proportionality factors,

E_{ijk} (E'_{ijk}) are base year (forecast) emissions,

b_j are background concentrations,

i indexes regions,

j indexes pollutants, and

k indexes sources.

Source-receptor proportionality factors are estimated by

$$P_{ijk} = D_{ijk} G_{ijk} \qquad (B.2)$$

where D_{ijk} are the percentage contributions of source k to exposure of pollutant j from all sources in region i; and

G_{ijk} are factors measuring relative frequency of stagnant air conditions and relative nearness of receptors to sources. Higher values for G_{ijk} reflect higher stagnation frequencies and less distance between sources and receptors.[1]

Average per-capita exposure to air pollution is estimated by

$$X'_{ij} = \sum_k P_{ijk} E'_{ijk}. \qquad (B.3)$$

Substituting this into equation B.1 and rearranging gives:

$$X'_{ij} = \frac{C'_{ij} - b_j}{C_{ij} - b_j} X_{ij}.$$ (B.4)

Thus forecast exposure equals base-year exposure scaled in proportion to the ratio of above-background, forecast concentrations to above-background, base-year concentrations. Exposures (and therefore damages) from man-originated activity are zero whenever forecast concentrations equal background concentrations.

Base-year measures of C_{ij} and b_j are from U.S. Environmental Protection Agency (1973) when regional data are available. For regions where base-year concentrations were not available to these regions above-background concentrations per weighted ton from known regions were assigned to these regions by Ridker and Watson (1978). The number of regions for which base-year data were available is as follows:

Pollutant	Number of AQCRs with Data Available
Particulates	172
Sulfur oxides	88
Nitrogen oxides	20
Carbon monoxide	57

There are 273 AQCRs (air quality control regions).

In the case of hydrocarbons, no regional concentration estimates are reported; therefore, equation B.1 could not be applied. The ambient-quality indexes reported for hydrocarbons are from equation B.3 with 1971 values set equal to 100. D_{ijk} is based on data in Krajewski et al. (1972) and Lewis (1971). G_{ijk} is based on Holzworth (1972).

Ambient water quality and average per-capita exposures were estimated by

$$PDI'_i = \frac{\sum_j \sum_k E'_{ijk}}{\sum_j \sum_k E_{ijk}} PDI_i$$ (B.5)

where $PDI_i (PDI'_i)$ are base-year (forecast) prevalence-duration-intensity index values.

Average per capita exposure to air pollution (as calculated by Eq. 4) was
used to estimate per capita base year damage for each region according to

$$\frac{AD_{ij}}{Pop_i} = \frac{(X_{ij}/m) \cdot \lambda_{ij}}{\sum_i Pop_i (X_{ij}/m)\lambda_{ij}} \; NAD_j \; ; \qquad (B.6)$$

where AD_{ij} is air damage in region i from pollutant j,

      Pop_i is population in region i,

      NAD_j is national damage from pollutant j,

      m is the median exposure over all regions in the base
year, and

      λ_{ij} are scalars related to X_{ij}/m in such a way as to
make damages a concave upward function of
exposure.

If λ_{ij} equaled 1 for all values of X_{ij}/m, then linear equation B.6 would
assign national damages to region i in proportion to its share of total
exposures. In comparison, a strictly concave upward relationship assigns
proportionately more national damages to regions with high exposures and
proportionately fewer damages to less polluted areas.

From equation B.6, a regional function for forecasting future year
damage is derived as follows:

$$\frac{AD'_{ij}}{Pop_i} = (X'_{ij}/m)\lambda_{ij} \; MD \qquad (B.7)$$

where MD are damages per capita in the base year for the
region with the median exposure.

Equation B.7 is derived from equation B.6 in the following way. In the base
year $(X_{ij}/m)\lambda_{ij}$ for the region with the median exposure equals 1.
Therefore, its damages per capita are

$$\frac{AD_{ij}}{Pop_i} = \frac{NAD_j}{\sum_i Pop_i (X_{ij}/m)\lambda_{ij}} . \qquad (B.8)$$

Substituting equation B.8 into equation B.7 gives:

$$\frac{AD'_{ij}}{Pop'_i} = (X'_{ij}/m)\lambda_{ij}\,\frac{NAD_j}{\sum_i Pop_i(X_{ij}/m)\lambda_{ij}}\,.\qquad\text{(B.9)}$$

That is, forecast damages per capita equal forecast exposures (normalized and scaled) multiplied by damages per exposure for the base year. By using equation B.7 (or its equivalent, equation B.9) and summing, it can be seen that damages over all regions for the base year equal national damages. Also, equation B.7 (by using a range of values for X_{ij}) gives linear segments whose slopes (relative to the median exposure value) are the same as those of a few selected empirical damage functions.

One additional adjustment to equation B.7 is made for years beyond the base year. An income-environment elasticity multiplier is applied:

$$\frac{AD'_{ij}}{Pop'_i} = (X'_{ij}/m)\lambda_{ij}\cdot MD\cdot Z' \qquad\text{(B.10)}$$

where

$$Z' = [(DI_t/DI_{t-1} - 1)L_t + 1]Z_{t-1}$$

DI is average national disposable income per income, and

L is the demand elasticity of environmental quality taken with respect to disposable income per capita; that is, a factor applied against the percentage change in disposable income to reflect changing valuation of the environment with respect to per-capita income.

Similar procedures are used to calculate regional and national water-pollution damages. In this case, PDI_i (as calculated by equation B.5) is the estimated exposure.

Note

1. For some sources and pollutants, G_{ijk} are reduced over time on the option that sources and receptors relocate away from each other.

References

Ahern, W.R. 1974. "Measuring the Health Effects of Reductions in Automotive Air Pollution." In *Federal Policy on Automotive Emissions Control,* eds. Jacoby, H., et al. Cambridge, Mass.: Ballinger.

Almon, Clopper, Jr., and Buckler, Margaret B., 1974. *1985: Interindustry Forecasts of the American Economy.* Lexington, Mass.: Lexington Books, D.C. Heath and Co.

American Conference of Governmental Industrial Hygienists. 1978. *TLVS: Threshold Limit Values for Chemical Substances in Workroom Air Adopted by ACGIH for 1978.* Cincinnati, Ohio: American Conference of Governmental Industrial Hygienists.

Ballou, Steve, et al. 1978. "Environmental Residual and Capital Cost Evaluation for Energy Recovery from Municipal Sludge and Feedlot Manure." Draft Report. Argonne, Ill.: Argonne National Laboratory, August.

Baumol, William, and Oates, W. 1975. *The Theory of Environmental Policy.* Englewood Cliffs, N.J.: Prentice-Hall.

Bechtel Corporation. 1975. *The Energy Supply Planning Model.* San Francisco, Calif.: August; NTIS PB-245 382, August.

Bennett, F.W., et al. 1973. *Processes, Procedures, and Methods to Control Pollution Resulting from Silvicultural Activities.* Kansas City, Mo.: Midwest Research Institute, U.S. Environmental Protection Agency, EPA 430/9-73-010, October.

Black & Veatch Consulting Engineers. 1978. "Environmental Assessment of Wind Turbine Power Plants." Preliminary Draft. March. Palo Alto, Calif.: Electric Power Research Institute, EPRI Project RP 955-1.

Boeing Engineering and Construction Co. 1977. *Central Receiver Solar Thermal Power System. Pilot Plant Preliminary Design Report. Vol. III. Collector Subsystem.* Washington, D.C.: U.S. Energy Research and Development Administration, SAN/1111-8/2, 29 April.

Booz, Allen, and Hamilton. 1975. *Strategic Environmental Assessment System, Executive Summary.* Prepared for the Environmental Protection Agency, Bethesda, Md., September 19.

Brumleve, T.D. 1977. *Eye Hazard and Glint Evaluation for the 5 MWth Solar Thermal Test Facility.* Albuquerque, N.M.: Sandia Laboratories, SAND 76-8022, May.

Caputo, R. 1977. *An Initial Comparative Assessment of Orbital and Terrestrial Central Power Systems: Final Report.* Pasadena, Calif.: Jet Propulsion Laboratory, No. 900-780, March.

Consroe, T.J.; Glaser, F.M.; and Shaw, R.W., Jr. 1976. *Potential Environ-*

mental Impacts of Solar Heating and Cooling Systems. Washington, D.C.: Environmental Protection Agency, Report no. EPA 600/7-76-014, October.

Council of Economic Advisors. 1975. *Economic Report of the President,* pp. 249–250. Washington, D.C.: U.S. Government Printing Office, February.

Davidson, M., and Grether, D. 1977. *The Central Receiver Power Plant: An Environmental, Ecological, and Socioeconomic Analysis.* Berkeley, Calif.: Lawrence Berkeley Laboratory, June.

Davidson, M.: Grether, D.; and Horowitz, M. *Assessment of the Socio-Economic and Environmental Aspects of the Central Receiver Power Plants.* Berkeley, Calif.: Lawrence Berkeley Laboratory.

Davidson, M.; Grether, D.; and Wilcox, K. 1977. *Ecological Consideration of the Solar Alternative.* Berkeley, Calif.: Lawrence Berkeley Laboratory, LBL-5927, February.

Dornbusch, D.M. 1973. *Benefit of Water Pollution Control on Property Values.* Report to the U.S. EPA, Washington, D.C.

Dunwoody, J.E. 1978. Resolving the Environmental Issues in Developing Fuels from Biomass. *Proceedings, Environmental Aspects of Non-Conventional Energy Resources-II,* 26–27 September 1978. Denver, Colo.: American Nuclear Society Topical Meeting.

Edelson, E., and Lee, T. 1976. *Preliminary Analysis of Industrial Growth and the Factors that Affect Industrial Growth Rates.* Pasadena, Calif.: Jet Propulsion Laboratory.

Energy and Environmental Analysis, Inc. 1977. *Draft Environmental Impact Statement of the National Solar Heating and Cooling Program.* Washington, D.C.: EEA. Environmental and Resource Assessment Branch, U.S. Department of Energy, 1 December.

Energy Information Administration. 1978. *Annual Report to Congress, Vol. II: Projections of Energy Supply and Demand and Their Impacts.* Washington, D.C. EIA. DOE/EIA-003612, April.

EPRI Journal. 1978. CO_2 and Spaceship Earth. 3: July/August.

Fann, J.C.C. 1978. Solar Cells: Plugging into the Sun. *Technology Review* 80, no. 8: 2–19.

Fisher, A., 1979. Assessing the Economic Effects of Implementing Air Quality Management Plans in California. Submitted to the California Air Resources Board.

Fisher, T.F., et al. 1976. Clean Fuels from Biomass, Sewage, Urban Refuse, and Agricultural Wastes. Presented at Symposium on Clean Fuels from Biomass, at Institute of Gas Technology, Orlando, Florida.

Flaim, Silvio, et al. 1978. *Economic Feasibility and Market Readiness of Eight Solar Technologies.* Golden, Colo.: Solar Energy Research Institute, SERI/TR-52-055d, September.

Fleischer, M. 1974. Environmental Impact of Cadmium: A Review by the Panel on Hazardous Trace Substance. *Environmental Health Perspectives* 253–323, May.

Garate, J.A., 1977. *Wind Energy Mission Analysis: Final Report.* General Electric Space Division, no. COO/2578-1/2, February.

Gauthier, J. 1978. *Environmental Assessment of Solar Energy-TASE, Preliminary Draft.* Oak Ridge, Tenn.: TRW, August.

Gianessi, L.P., et al. 1977. *The Distributional Implications of National Air Pollution Damage Estimates,* Reprint 150. Washington, D.C.: Resources for the Future, Inc.

Gikis, B.J., et al. 1977. *Preliminary Environmental Assessment of Energy Conversion Processes for Agricultural and Forest Product Residues.* Vol. I, Draft Final Report. Stanford, Calif.: Stanford Research Institute.

Hall, E.H., et al. 1976. *Comparison of Fossil and Wood Fuels.* Washington, D.C.: U.S. Environmental Protection Agency, EPA-600/2-76-056, March.

Harris, M.T., Jr., et al. 1968. The Residence Site Choice, *The Review of Economics and Statistics* 60: 241–247.

Hashimoto, A.G.; Chen, Y.R.; and Prior, R.L. 1978. Methane and Protein Production from Animal Feedlot Wastes. In *Proceedings of the 33rd Annual Meeting, Soil Conservation Society of America.* 30 July–August 1978, Denver, Colo.

Heintz, M.T., Jr., et al. 1976. *National Damages of Air and Water Pollution.* Washington, D.C.: U.S. Environmental Protection Agency, September.

Holzworth, G.C. 1972. *Mixing Heights, Wind Speeds, and Potential for Urban Air Pollution Throughout the Contiguous United States.* Raleigh, N.C.: U.S. Environmental Protection Agency; AP-101.

House, Peter. 1977. *Trading Off Environment, Economics, and Energy.* Lexington, Mass.: Lexington Books, D.C. Heath and Co.

Howlett, K., and Gamache, A. 1977. *Silvicultural Biomass Farms. Vol. VI. Forest and Mill Residues as Potential Sources of Biomass.* Energy Research and Development Administration, MTR-7347, May.

Hyde, J. 1978. Environmental Assessment of Solar Energy-Space Heating and Cooling. Memo to John Altseimer. Los Alamos, N.M.: Los Alamos Scientific Laboratory, July 27.

Krajewski, E.P. et al. 1972. *A Study of the Relationship between Pollutant Emissions from Stationary Sources and Ground-Level Ambient Air Quality.* Washington, D.C.: U.S. Environmental Protection Agency.

Krawiec, Frank. 1979. *Economic Measurement of Environmental Damages: A Review of Literature.* Golden, Colo.: Solar Energy Research Institute, SERI/PR-52-311, September.

Lake, E., et al. 1976. *Classification of American Cities for Case Study Analysis,* Vol. III. Report to the U.S. Environmental Protection Agency, Cambridge, Mass.

Lawrence, Kathryn A. 1979. *A Review of the Environmental Effects and Benefits of Selected Solar Energy Technologies.* Golden, Colo.: Solar Energy Research Institute, SERI/TP-53-144R, May.

Lerner, William. 1972. *City and County Data Book.* Washington, D.C.: U.S. Department of Commerce.

Lewis, D.M. 1971. *Allocations of Air Pollution Control Research Found on the Basis of Human Experience.* Washington, D.C.: U.S. Environmental Protection Agency.

Lindstrom, M.J., et al. 1978. Soil Conservation Limitations on Removal of Crop Residues for Energy Production. In *Proceedings of the 33rd Annual Meeting: Soil Conservation Society of America.* 30 July–2 August 1978, Denver, Colo.

Marcuse, William. 1978. *Internalizing the Externalities of Solar Technology: Methodologies for Incorporating Externalities in the Assessment of Policy Options and Technology Assessments of Solar Energy Initiatives and R&D Programs Using Brookhaven Models.* Upton, N.Y.: Brookhaven National Laboratory, May.

Martin Marietta Corp. 1977. *Central Receiver Solar Thermal Power System, Phase I, Preliminary Design Report, Draft,* Vol. VII. Denver, Colo.: MCR-77-156, June.

Mendelson, R., and Orcott, G. 1979. An Empirical Analysis of Air Pollution Dose-Response Curves. *Journal of Environmental Economics and Management* 6: 85–106.

Metzger, C.N. 1977. *Inforum Targeting Procedure.* Prepared for U.S. Department of Energy and Environmental Protection Agency. Rockville, Md.: Control Data Corp., September.

Mierynck, W.H. 1965. *The Elements of Input-Output Analysis.* New York, N.Y.: Random House.

MITRE Corp. 1975. *Preliminary Environmental Assessment Concerning the Construction and Operation of a 5-MW Solar Thermal Central Receiver Test Facility.* McLean, Va.: MITRE Corp., 28 November.

MITRE Corp., Metrek Div. 1977. *Systems Descriptions and Engineering Costs for Solar-Related Technologies. Vol. II: Solar Heating and Cooling of Buildings (SHACOB).* U.S. Energy Research and Development Administration, MTR-4785, June.

MITRE Corp. 1977a. Annual Environmental Analysis Report: A Preliminary Environmental Analysis of Energy Technologies Using the Assumptions of the National Energy Plan, Vol. IV, Simulation Data Base. Draft. McLean, Va., 6 September.

―――. 1977b. *Systems Descriptions and Engineering Costs of Solar-Related Technologies.* McLean, Va., June.

———. 1977c. *Solar Energy: A Comparative Analysis to the Year 2020, The SPURR Methodology*. McLean, Va., September.

———. 1977d. *The Need for Power Studies: An Assessment of the Adequacy of Future Electric Generating Capacity*. McLean, Va.: MTR 7549, July.

———. 1977e. *Silvicultural Biomass Farms, Vol. II: Land Suitability and Availability*. McLean, Va.: Technical Report 7347.

———. 1978a. *A Short Course on the SEAS Model*. McLean, Va., May.

———. 1978b. "Estimates of Solar Savings," Scenario: Base in MITRE memorandum in "Documentation of Baseline Scenarios for the DPR." Memorandum no. W52-M-269, 25 July 1978.

National Academy of Sciences. 1977. *Report of the Solar Resource Group, Committee on Nuclear and Alternative Energy Systems (CONAES)*. Washington, D.C.: National Academy of Sciences, February.

. 1977. *Medical and Biologic Effects of Environmental Pollutants. Nitrogen Oxides*. Washington, D.C.: National Academy of Sciences.

Neenan, B., et al. 1979. *Cost and Environmentaal Data for Selected Solar and Non-solar Technology Applications*. Los Alamos, N.M.: Los Alamos Scientific Laboratory.

Olsen, N.A., et al. 1973. *Environmental Aspects of Cadmium Sulfide Usage in Solar Energy Conversion—Part I, Toxicological and Environmental Health Considerations—A Bibliography*. Washington, D.C.: National Science Foundation, NTIS Report No. PB 238-285, 1 June.

Povich, M.J. 1977. Fuel Farming-Water and Nutrient Limitations. In *Proceedings, Second Pacific Chemical Engineering Congress*. 23–31 August 1977, Denver, Colo., pp. 743–747.

Ridker, R.G., and Watson, W.D. 1978. *To Choose a Future: Resource and Environmental Problems of the U.S., A Long-Term Global Outlook*. Washington, D.C.: Resources for the Future, Inc., Appendix A.

Rogers, S.E., et al. 1976. *Evaluation of the Potential Environmental Effects of Wind Energy System Development*. Washington, D.C.: U.S. Energy Research and Development Administration, ERDA/NSF/ 07378-75/1, August.

Roop, R.D. 1978. *Energy from Biomass: An Overview of Environmental Aspects*. Presented at the 2nd National Conference on Technology for Energy Conservation, 24–27 January 1978. Albuquerque, N.M.: Oak Ridge National Laboratory, Publication no. 1140.

Searcy, J.Q., ed. 1978. *Hazardous Properties and Environmental Effects of Materials Used in Solar Heating and Cooling (SHAC) Technologies: Interim Handbook*. Albuquerque, N.M.: Sandia Laboratories, SAND 78-0842, August.

Sears, D.R., and McCormick, P.O. 1977. *Preliminary Environmental Assessment of Solar Energy Systems*. Lockheed Missiles and Space Co., Inc., EPA-600/7-77-086, August.

Sears, D.R.; Merrified, D.V.; and Perry, M.M. 1977. *Environmental Impact Statement for a Hypothetical 1000 MWe Photovoltaic Solar-Electric Plant.* Lockheed Missiles and Space Co., Inc., EPA-600/7-77-085, August.

Sessler, G., and Cukor, P.M. 1975. *Pollutant Releases, Resource Requirements, Costs, and Efficiencies of Selected New Energy Technologies.* Teknekron, Inc., December.

Sittig, M. 1975. *Environmental Sources and Emissions Handbook.* Park Ridge, N.J.: Noges Data Corp.

Stanford Research Institute. 1976. *An Evaluation of the Use of Agricultural Residues as an Energy Feedstock,* Vol. II. Palo Alto, Calif.: SRI Project 3520.

———. 1977. *Solar Energy in America's Future.* Palo Alto, Calif.: Stanford Research Institute, DSE-115/1, March.

Torkelson, L.E., et al. 1978. *A Summary of Current Solar Collector Cost and Performance Data, Internal Report.* Albuquerque, N.M.: Sandia Laboratory, 1 March.

Truett, J.B., et al. 1975. Development of Water Quality Management Indices. *Water Resources Bulletin.* II (no. 3): June.

U.S. Department of Commerce. 1974. *OBERS Projectional Economic Activity in the U.S.,* vol. IV. Washington, D.C.: U.S. Department of Commerce.

———. 1977. *Forecast of Likely U.S. Energy Supply/Demand Balances for 1985 and 2000 and Implications for U.S. Energy Policy.* Washington, D.C.: U.S. Department of Commerce, PB-266240, 20 January.

U.S. Department of Energy. 1978a. *Environmental Development Plan: Solar Heating and Cooling of Buildings, 1977.* Washington, D.C.: U.S. Department of Energy, DOE/EDP-0001, March.

———. 1978b. *Environmental Development Plan: Solar Agricultural and Industrial Process Heat, 1977.* Washington, D.C.: U.S. Department of Energy, DOE/EDP-0002, March.

———. 1978c. *Environmental Development Plan: Solar Thermal Power Systems, 1977.* Washington, D.C.: U.S. Department of Energy, DOE/EDP-0004, March.

———. 1978d. *Environmental Development Plan: Wind Energy Conversion, 1977.* Washington, D.C.: U.S. Department of Energy, DOE/EDP-0007, March.

———. 1978e. *Environmental Development Plan: Photovoltaics, 1977.* Washington, D.C.: U.S. Department of Energy, DOE/EDP-0003.

U.S. Environmental Protection Agency. 1973. *The National Air Monitoring Program: Air Quality and Emissions Trends.* Research Triangle Park, N.C.: Environmental Protection Agency, EPA-450/1-73-001-b.

———. 1977. *Compilation of Air Pollutant Emission Factors,* 3rd ed., supplements 1–7, parts A and B. Washington, D.C.: U.S. Environmental Protection Agency, AP-42, May.

———. 1978. *Compilation of Air Pollutant Emission Factors,* supplement 8. Washington, D.C.: U.S. Environmental Protection Agency, AP-42, May.

U.S. Energy Research and Development Administration. 1977a. *Solar Program Assessment: Environmental Factors; Solar Heating and Cooling of Buildings.* Washington, D.C.: U.S. Energy Research and Development Administration, ERDA 77-47/1a, March.

———. 1977b. *Solar Program Assessment: Environmental Factors; Photovoltaics.* Washington, D.C.: U.S. Energy Research and Development Administration, ERDA 77-47/3, March.

———. 1977c. *Solar Program Assessment: Environmental Factors; Solar Thermal Electric.* Washington, D.C.: U.S. Energy Research and Development Administration, ERDA 77-47/4, March.

———. 1977d. *Solar Program Assessment: Environmental Factors: Wind Energy Conversion Systems (WECS).* Washington, D.C.: U.S. Energy Research and Development Administration, ERDA 77-47/6, March.

———. 1977e. The Need for Deployment of Inexhaustible Energy Resource Technologies, Report of Inexhaustible Energy Resources Planning Study (IERPS), Draft Final Report. Washington, D.C.: U.S. Energy Research and Development Administration, March.

———. 1977f. *Solar Program Assessment: Environmental Factors; Fuels from Biomass.* Washington, D.C.: U.S. Energy Research and Development Administration, ERDA 77-47/7, March.

———. 1978. Market Oriented Program Planning Study (MOPPS), Draft Final Report. Washington, D.C.: U.S. Energy Research and Development Administration, March.

Vermont Castings, Inc. 1976. *Operation Manual for the Defiant and Vigilant Woodburning Parlor Stoves.* Randolph, Vt.: Vermont Castings, Inc.

Windholz, M., ed. 1976. *The Merck Index: Encyclopedia of Chemicals and Drugs,* 9th ed. Rahway, N.J.: Merck and Co., Inc.

Yokell, M.D. 1978. *Economic Measurement of Energy-Related Environmental Damages, A Workshop Summary.* Golden, Colo.: Solar Energy Research Institute, SERI/TP-52-058, December.

Zalinger, Stein. Omnium-G Company, Anaheim, Calif., July 1978. Telephone communication to Harit Trivedi.

Index

ABATE, 38
Acurex 3001 Model, 40
Aggregated subarea, 27
Agricultural and forest residue
   collection, 42, 113–114
Agricultural and industrial process heat,
   1, 21, 29, 33, 39–40, 106
Ahern, W.R., 37
AIPH. *See* Agricultural and industrial
   process heat
Air-quality-control regions, 20, 27
Almon, Clopper, Jr., 15
A matrix, 6, 15–20
AMBIENT, 5, 7–8, 9, 34, 127–130
American Conference of Governmental
   Industrial Hygienists, 110
Anaerobic digestion of manure, 42,
   114–115

Ballou, Steve, 42, 114, 115
Base case scenarios, 5, 24, 43–49, 58–64
Baumol, William, and W. Oates, 2
Bechtel Corporation, 15
BENEFITS, 5, 8, 9–10, 35–38, 67, 68,
   123, 127–130
Bennett, F.W., 112, 113, 114
Best Practical Technology regulations,
   26
Bio Gas of Colorado, Inc., 42
Biomass for energy, 1, 22, 33, 42–43
Birds, 111
Black and Veatch Consulting Engineers,
   112
B matrix, 15
Boeing Engineering and Construction
   Co., 109
Booz, Allen, and Hamilton, 14
Bottom-up regional aggregation, 9
Brookhaven National Laboratory, 13
Brumleve, T.D., 109, 111

Cadmium, 109–110
Caputo, R., 107

Carbon dioxide, 64–67
Clean Air Act, 26
Clean Water Act, 26
CONAES, 23
Consroe, T.J., 105

Davidson, M., 107, 108, 114
Defiant stove, 43
Desert sites, 107–108
Domestic Policy Council, 119
Domestic Policy Review, 22–25, 43–45
   passim
Dornbusch, D.M., 37
DPR. *See* Domestic Policy Review
Dunwoody, J.E., 112, 113

Edelson, E., and T. Lee, 55
EIM. *See* Energy investment module
Electric utilities module, 30
Energy and Environmental Analysis,
   Inc., 104, 105, 106
Energy Information Administration
   projections, 29, 30–31
Energy investment module, 14–15, 26,
   126
Energy System Network Simulator, 14,
   21–22
Energy targeting, 26–27
Environmental analysis, 5
Environmental comparisons, 57–67;
   economics of, 67–79
Environmental controls, 26, 63–64
Environmental issues, 100–118
Environmental Protection Agency, 26,
   109
ESNS. *See* Energy System Network
   Simulator

Fann, J.C.C., 109
Federal Aviation Administration, 111
Final demands, 6
Fisher, A., 10, 36, 37, 70
Fisher, T.F., 43

Flaim, Silvio, 2
Flat-plate AIPH hot-water system, 40
Fleischer, M., 110

Gauthier, J., 42
General Electric, 42
Gianessi, L.P., 36
Gikis, B.J., 116
GNP. *See* Gross national product
Gross national product, 6, 55, 70; GNP
    targeting, 27, 57–58

Hall, E.H., 112, 113, 116, 117
Harris, M.T., Jr., 38
Hashimoto, A.G., 115
Heintz, M.T., Jr., 36
Heliostats, 108–109
Hexel Corporation Parabolic Trough,
    40
High solar-energy-use scenarios, 24–25,
    49–55, 58–64
House, Peter, 14
Howlett, K., and A. Gamache, 113, 114
Hyde, J., 39

IERPS, 23
Income-environment elasticities, 38
Industrial Boiler Module, 31
INFORUM, 6–7, 8, 15–20, 33–34
Input-output analysis, 8, 11–12. *See also*
    INFORUM
INSIDE, 20
Intermediate demands, 6
Iterative national-regional scenario
    development process, 122–123

Kaman Aerospace Corporation, 42
Krawiec, Frank, 36

Lake, E., 35
LANDUSE, 21
Lerner, William, 29
Life-cycle emissions, compared, 96–99
Lifetime residuals, defined, 85
Lindstrom, M.J., 114
Linear programming energy models,
    13–14

Maier, Mike, 55
Manure collection, 42
Marcuse, William, 14
Market failure, 2
Martin Marietta Corp., 41
Maximum-feasible solar scenario, 5, 24,
    25, 49–53, 62–64
Maximum-practical solar scenario, 5,
    24, 25, 49, 58–64
Mendelson, R., and G. Orcott, 36
Metzger, C.N., 27
Miernyck, W.H., 12
MITRE, 14, 22, 29, 39, 40, 42, 43, 106
MITRE/SPURR, 23
MOPPS, 22

National-air-pollution-damage estimate,
    36
National Electric Reliability Council, 29
National Energy Plan, 31, 55
NEP. *See* National Energy Plan
New Source Performance Standards,
    26, 121

O & M *See* Operation and management
OBERS, 9, 20, 29
Ocean thermal-energy conversion, 2, 53
Olsen, N.A., 110
Omnium-G Solar-Powered Electrical
    Generating Plant, 40
Operation and management, 123
OTEC. *See* Ocean thermal-energy
    conversion

Parabolic-dish total-energy system, 40
Parabolic-trough AIPH steam system,
    40
PDI index, 35
Photovoltaics, 1, 21, 33, 41, 109–110
PIES. *See* Project Independence
    Evaluation System
Policy implications, 118–121
Positive externalities, 2
Povich, M.J., 112
Process analysis, 11
Project Independence Evaluation
    System, 30

Purox System, 43, 115–116
PV. *See* Photovoltaics
Pyrolysis of agricultural residues, 43,
    115–116

REGION, 8–9, 20
Regionalization of energy use, 27–31
Research, 121–124
RESGEN, 21, 31, 33–34
RFF. *See* Resources for the Future, Inc.
Ridker, R.G., and W.D. Watson, 33,
    34, 36, 37
Rogers, S.E., 111
Roop, R.D., 115

Scenario-based analysis, 5, 12–13, 57–79
Searcy, J.Q., 104
Sears, D.R., 108
Sears, D.R., and P.O. McCormick, 108
SEAS. *See* Strategic Environmental
    Assessment System
Sectors, 11–12
SERI scenarios, 24, 25, 30–31, 41, 43–55
Sessler, G., and P.M. Cukor, 108
SHACOB. *See* Solar heating and
    cooling of buildings
SIC. *See* Standard industrial
    classification
Silvicultural farms, 42, 112–113
SIP. *See* State implementation plan
Site-specific analysis, 121–122
Solar energy, defined, 48–49
Solar-energy scenarios, 5, 22–26, 43–55
Solar-energy systems, generic, 1
Solar-energy technologies, generic
    descriptions, 39–43
Solar heating and cooling of buildings,
    1, 21, 29, 33, 39, 102–106
Solar-power-satellite stations, 2, 53
Solar-residuals coefficients data, 125

Solar-thermal-electric system, 1, 21–22,
    33, 40–41, 107–109
Source-receptor transfer coefficients, 35
Spaghetti-bowl diagram, 6
SPSS. *See* Solar-power-satellite stations
SPURR, 23, 27
SRI. *See* Stanford Research Institute
Stand-alone analysis, 5, 12–13, 79–118
Standard industrial classification, 20
Stanford Research Institute, 22, 29
State implementation plan, 26
Steady-state condition, defined, 85
Strategic Environmental Assessment
    System, 5, 6–7, 8–9, 13–34, 79–118,
    123

Top-down regional disaggregation, 8–9
Torkelson, L.E., 40
Total demands, 6
Total-energy system, 21
TRANSPORTATION, 21
Truett, J.B., 35

Union Carbide Corporation, 43

Vermont Castings, Inc., 43

Watson, W.D., 33–38 passim
WECS. *See* Wind energy conversion
    system
Wind energy conversion system, 1, 22,
    33, 42, 110–112
Wind farms, 111
Wood-fired steam-electric power plant,
    43, 116–117
Wood stoves, 43
Workshop Summary, 10

Yokell, M.D., 10, 121, 123

Zalinger, Stein, 40

About the Author

Michael D. Yokell is a principal and chief administrative officer of Energy and Resource Consultants, Inc., an economic and technical analysis firm located in Boulder, Colorado, providing research and consulting services in the fields of energy, resource development, and environmental matters. Prior to founding Energy and Resource Consultants, Inc., Dr. Yokell was senior economist at the Solar Energy Research Institute, Golden, Colorado. He has been on the faculty of the University of California at Berkeley, Washington State University, and Colorado State University.

Dr. Yokell received the B.Sc. in physics from MIT and the Ph.D. in economics from the University of Colorado. He is the author of many articles and one book, *Yellowcake: The International Uranium Cartel,* and is the senior author and editor of a forthcoming textbook on the economics of solar and alternative energy systems. Dr. Yokell is a specialist in the economics of solar and alternative energy systems, in uranium economics, in water economics, and in natural-resource problems affecting the western states.